ABANDONED & VANISHED CANALS OF IRELAND, SCOTLAND AND WALES

Andy Wood

Front cover top: Monkland Canal, used with permission of John Alexander Rae.
Front cover bottom: The Grand Canal in Dublin, used with permission of William Murphy.
Back cover: Lismore Canal, used by permission of Mark Morton.

First published 2015

Amberley Publishing
The Hill, Stroud
Gloucestershire, GL5 4EP

www.amberley-books.com

British Library Cataloguing in Publication Data.
A catalogue record for this book is available from the British Library.

ISBN 978 1 4456 4868 2 (print)
ISBN 978 1 4456 4869 9 (ebook)

Typeset in 10pt on 12pt Sabon.
Typesetting and Origination by Amberley Publishing.
Printed in the UK.

Contents

Introduction

There is nothing I love more than cruising gently and slowly on a canal boat. I also appreciate the sterling work done by the volunteers of the Waterways Recovery Group (WRG), the Inland Waterways Protection Society (IWPS) and the many other enthusiasts who put their hearts and souls – and muscle power – into restoring their local canals in many different parts of the British Isles.

One day, I idly wondered how many other canals and navigations had become disused or had disappeared altogether. As a writer and historian, it was inevitable that I would conceive of a book that would tell the stories of the ghostly waterways of the British Isles. As England alone has more lost waterways than Ireland, Scotland and Wales put together, my publisher decided that it merited a volume to itself. *Abandoned and Vanished Canals of England* was published in June 2014.

This second volume will, I hope, surprise the reader with the number of lost canals in Ireland, Scotland and Wales. Given that in the 'Canal Age' the Republic of Ireland was still more than a century away from becoming a reality, I feel it is appropriate to include the now lost or abandoned canals throughout the whole island of Ireland.

In Scotland, the canals that survive to the present day are the Forth & Clyde Canal and the Union Canal that were built to connect the major cities of the industrial Central Belt. They also provide a shortcut for boats to cross between the west and the east without a sea voyage. The Crinan Canal removed the need for a long diversion around the Mull of Kintyre, and the Caledonian Canal performed a similar function in the Highlands of Scotland. However, a canal that was intended to link Glasgow, Paisley and Johnstone directly to the west coast of Scotland never extended further than Johnstone, and so falls within the scope of this book. The Monkland Canal, which also features in this book, was conceived in 1769 by tobacco merchants and other entrepreneurs as a way of bringing cheap coal into Glasgow from the coalfields around Monkland. Having lost traffic to the railways, and suffering from problems with its water supply, it was legally abandoned for navigation in 1942, although it was not until 1954 that work began on filling it in. Today only fragments of the Monkland Canal remain.

The majority of Welsh canals were concentrated in South Wales. A number of isolated canals run along the South Wales valleys, including the Swansea Canal, the Neath & Tennant Canal, the Glamorganshire Canal and the Monmouthshire & Brecon Canal. Nearly all of these canals were constructed to serve local industries, principally the coal and iron industries, and fell derelict when faced with competition from railways.

Of course, as with *Abandoned and Vanished Canals of England*, I am the first to acknowledge that I stand on the shoulders of giants, some living and some who have embarked on their final cruise. Among them are Joseph Priestley, Henry Rodolph de Salis, Edward W. Paget-Tomlinson, L. T. C. (Tom) Rolt, Charles Hadfield, Joseph

Boughey (a fellow member of the Boat Museum Society at the National Waterways Museum, Ellesmere Port), Ronald Russell, P. J. G. (John) Ransom, Anthony Burton, D. D. Gladwin, Brian J. Goggin, W. A. McCutcheon, and Jane Cumberlidge.

IRELAND
Athlone Canal

The cutting of a canal, 1 mile (1.6 kilometres) long, on the west side of the town of Athlone in 1757 was the start of one of the earliest artificial inland waterways in Westmeath. The work was necessary as the River Shannon was not navigable through Athlone because of rapids and shallows, which formed a natural ford for crossing the water but were an obstruction to boats.

Today, boats passing through Athlone use a lock in the river next to the weir and downstream of the road bridge. The lock, weir and bridge were constructed by the Shannon Navigation Commissioners in the 1840s. In the past, however, boats used a canal, about 1.5 miles (2.4 kilometres) long, west of the river.

The Athlone Canal was built by Thomas Omer for the Commissioners of Inland Navigation. Omer is thought to have been Dutch, but he had worked on a number of river navigations in England before being invited to Ireland in 1755. Work on the Athlone Canal began in 1757 and 325 men were employed in a labour scheme for the poor. It cost £30,000 to cut the canal.

Omer built a single lock measuring 120 feet by 19 feet (36.5 by 5.7 metres), with a rise of 4 feet 6 inches (1.3 metres), and a guard lock, further upstream, with a single set of gates to protect the canal against floods. There were two small harbours or quays for boats waiting to use the lock, one above and one below the lock. The canal was one of the first attempts to make the River Shannon fully navigable at Athlone, although it was not until the mid-nineteenth century that this actually happened with the completion of the Shannon Navigation works. By 1849 the Athlone Canal was obsolete.

In September 2011 there was a public meeting about the proposed restoration of the canal at the Prince of Wales Hotel, Athlone. The meeting heard from the Inland Waterways Association of Ireland, the Boyne Canal Restoration Group, who explained the history of the canal, and an environmentalist who spoke about the environmental benefits of restoration.

A local councillor said that, at about 1.25 miles (2 kilometres) long, the canal was an infant in terms of canals compared with the Shannon–Erne Navigation, which was 40.5 miles (65 kilometres) long, and the Boyne, which was 12.5 miles (20 kilometres) long – both of which had undergone restoration schemes in recent years.

He said that Waterways Ireland (the Irish equivalent of the Canal & River Trust in the UK) would be the key to making the restoration happen as it was responsible for all of the navigable sections of the river. The stretch from Battery Bridge to Mick McQuaid's Bridge presented its own challenge because 25 yards (22.8 metres) of it were infilled and were now the site of a children's playground.

The canal is no longer navigable, but small boats are moored at its upper end and the entire line can easily be followed. At a meeting of Athlone Town Council in March

2013, the councillors heard a proposal to re-water the infilled section of the canal and to provide a boardwalk. It was said that planting hedging and building a fence along the length of the canal would create public awareness and encourage people to walk along its route. In general, the councillors welcomed the plan, although it was pointed out that there was no funding for it in that year's budget.

Ordnance Survey of Ireland Grid Reference: 603369 741343

Ballinasloe Canal

Ballinasloe is the largest town in County Galway. The Guinness company used the town's canal stores to store and distribute stout to the Irish Midlands. The Grand Canal allowed the Guinness barges to travel easily from Dublin to Shannon Harbour.

The River Shannon itself was made navigable in the mid-eighteenth century. Ballinasloe was linked to it by its own canal in the 1820s. To call the canal to Ballinasloe a branch of the Grand Canal was inaccurate, since it did not actually connect directly with the Grand Canal. The main line of the Grand Canal reached Shannon Harbour in 1804, connecting with the Shannon on its east bank. Some twenty years later, the Ballinasloe Canal was opened, branching off from the west bank of the river.

The canal closed in 1961 and most of its route is used today by a Bord Na Mona (BNM) light railway, which is used in connection with turf extraction. Along the route of the canal, the Kylemore Lock still exists, with its gates and cill removed, through which the BNM railway runs. The walls, gate recesses and bridge all remain.

In 1978 the old Canal House, built as offices for the canal company in Ballinasloe, was purchased by the Department of Justice to be used as a training centre for young people who were before the courts or at risk in the community.

In the 1990s, the River Suck, the main tributary of the River Shannon and which meets the Shannon a few miles north of the village of Shannonbridge, was made navigable from Ballinasloe to its confluence with the Shannon, just south of Shannonbridge.

Ordnance Survey of Ireland Grid Reference: 583985 731803

An original lock gate of the Ballinasloe Canal, still in place at Kylemore in County Galway. Image used by permission of Ted McAvoy.

Bridgetown Canal

The Bridgetown Canal, 5 miles (8 kilometres) long, was built between 1850 and 1853 and served the communities of Bridgetown and Duncormick. It was made necessary by the building of the Cull Bank on the south coast of County Wexford, bordering the Atlantic Ocean, which prevented access to the rivers of the area. The eastern part of the shallow, tidal inlet of the sea and the estuary of the Duncormick River, otherwise known as the Cull, was reclaimed in the nineteenth century by the Cull Bank and is now polder-like land, farmed for arable crops and grassland.

In 1844, a report was published on the 'Proposed Drainage of the Flooded Lands in the District of Ballyteigue'. A schedule attached to the report gave a detailed description of 'cutting and sinking a navigable canal from the tideway at Blackstone to the Village of Bridgetown'. The description of this canal is different to the one given in a later schedule of 1888 of 'a new main channel, cut for a length of about two statute miles (3.2 kilometres) from Killag River Bridge to the outlet at level of sea; low water springs at Crossfarnoge, Kilmore'.

At first, the canal was intended simply as part of a drainage and land reclamation scheme, but then it was realised that it would be useful for boat traffic. There were two lengths – the Bridgetown Canal proper and a branch to Ballyteigue. There was still a reasonable amount of regular traffic at the time of the First World War, and the canal remained in use up to the 1940s.

In the 1830s, the village of Duncormick had a population of about 250, and there was a steady trade in imported slate, coal, limestone and culm from South Wales. The 1837 *Topographical Dictionary of Ireland* said:

> Vessels of 100 tons [101.6 tonnes] burden could cross the bar [the sandbar between open sea and the inner water] and lie securely in the Lough. This was no longer possible once the lake of Ballyteigue, or Lough Tay, was drained bringing 1,630 acres [659.6 hectares] into use as farmland, hence the decision to construct the canal.

The Lough area had been known locally as the 'Little Say' and the entrance was usually called the Bar of Lough, a term that was also used in a document from the Commissioners of Public Works, which mentions 'Coal Yards at Lough, Lacken and Duncormick'. In 1844, Mr Joshua Lett, Officer of the Coastguards and Customs at the Bar of Lough, reported that forty-seven ships had entered the Lough between March and September of that year.

In 1979, Paddy Redmond, who had lived in the area all his life, remembered two of the ships, two-masted schooners about 80 feet (24.3 metres) long, which sailed from the area until 1913. He spoke of how important water transport was to the area, and said,

… the cargoes could be anything; sometimes they took potatoes or grain to Wales and brought culm, coal dust, and dirt back with them, and the brick for building the railway bridge at the Mill O'Rags came into Duncormick by lighter after being off-loaded from a larger vessel at Lough.

The canal certainly earned its keep, allowing small vessels to carry kelp, sand and coal to Bridgetown, improving the supply of water to two water-mills, and acting as a mainland drain for more than 1,000 acres (404.6 hectares) which had previously frequently flooded. The scheme for building the canal had been instigated by Mr Rowe, a landowner at Ballycross. In the first instance it was a relief measure for the poor, who were being affected by the famines of the early and mid-nineteenth century, as much as it was a consideration of its usefulness carrying goods. It was certainly the case, possibly due to the construction of the canal and other factors, that the population of Kilmore actually increased during the famine years.

The report of the Commissioners of Public Works refers to the building of 'Salt Bridge, Liffey Bridge, Redmoor Bridge, Yoletown Bridge … erecting sluices, cutting a navigable branch canal at Pembrokestown and at Newhouse'. It also describes building a 'Quay at the Village of Bridgetown' and 'Making two inclined loading and landing slips in canal banks, one at Gibberpatrick and one at Sheephouse'.

Folk memories of the building of the canal remain today. Initially, there were strong objections to the drainage scheme for the area from fishermen, landowners and merchants who made their living from the lough. Nevertheless, the work began under the management of the Board of Public Works. The drainage plan involved building a wall with four sluice gates across the lough at the Cull, the excavation of the canal to Bridgetown to drain water from the Mayglass and Bridgetown area, and a bridge to connect Ballyburn to Killag, which came to be known as the 'Big Bridge'. Four quays were built at Seafield, Redmoor, Yoletown and Bridgetown, with the latter being known locally as 'New Quay'.

The work finished in 1853, and Paddy Duffin was appointed as caretaker at the Cull, his duties being to open the sluice gates on the falling tide and to keep them free of seaweed.

The labour force involved in building the canal and the other works was enormous: thousands of men were employed, each receiving 3*d*, and 1 pound (0.45 kilograms) of maize per day. Men would walk up to 12 miles (19.3 kilometres) to work on the drainage scheme every day, walking home again when they had finished work.

The little stone bridges over the Bridgetown Canal were well-designed and constructed. Particularly impressive was a skew bridge, which required sound knowledge and confidence in its design and construction. It seems possible that the engineer responsible for the canal was William Chapman, who went to Ireland in the late eighteenth century, initially as an agent for Boulton & Watt steam engines, and stayed to make a career in canal engineering.

The construction of the usual hump-backed canal bridges, which took roads over the canal, was a straightforward job, but crossing at an angle would have presented problems to an inexperienced engineer. Chapman, however, had already been the first to build skewed hump-backed bridges in the British Isles. The bridges over the Bridgetown Canal are built of cut local limestone, with elegantly shaped, domed capstones. They were constructed with such skill and soundness that they have survived for some 180 years. The Bridgetown Canal as a whole was a busy and valuable waterway for nearly ninety years, not quite falling out of use until 1940.

It is said that, to this day, the canal is the chosen route of one local farmer who travels by boat along the canal to enjoy his pint of stout in a pub in Duncormick. It would not be surprising if the pub that draws him is Sinnott's on the corner of Duncormick village, stepping inside which is like stepping back in time, among the ghosts of the men who built the Bridgetown Canal.

Ordnance Survey of Ireland Grid Reference: 698861 609762

Broadstone Branch (Royal Canal)

The name Broadstone Branch suggests that it was a less important canal than the Royal Canal Docks in Dublin. In 1790, the Lord Lieutenant of Ireland, the Earl of Westmoreland, laid the first stone of a lock at Phibsborough, at what is now called Cross Guns Bridge. At the time, the lock and the bridge were both named in honour of the earl. The Ordnance Survey of Ireland (OSI) map of 1837 shows the lock at Phibsborough as the fifth, with four more (Nos 4, 3, 2 and 1) between that point and the Royal Canal Docks, and a sea lock between there and the Liffey.

The terminal basin of the Royal Canal, as originally planned, was not at the Liffey but at the Broadstone, one of the older parts of the city, which was known in earlier times as Glasmanogue. In evidence to the Railway Commissioners in 1838, the Royal Canal Company Engineer, Mr Tarrant, said that the canal had two branches, one of which went to Longford, while of the other he stated the 'branch from the Broadstone level is 2 miles (3.2 kilometres), when it enters the River Liffey'. Although the branch was completed by 1796, it was another ten years before the harbour was opened, and construction continued for some years after that. The line to the Liffey was clearly of less importance, as the harbour at the Broadstone was close to the city markets, the Linen Hall and the King's Inns. There was a Royal Canal Hotel at the Broadstone, from where the passenger-carrying 'passage-boats' left.

In an engraving of Broadstone dating from 1821, the most prominent building is the King's Inns, designed by James Gandon in 1800 and, by 1821, nearing completion. The Inns were the home of the Honourable Society of King's Inns, the Irish equivalent of the Inns of Court in London. The society controlled the entry of barristers into the justice system of Ireland. The building looks much the same today, but the real surprise on viewing the engraving is to find a harbour immediately in front of the King's Inns, because it has long since been infilled.

The Royal Canal Company soon realised, however, that more money was to be forthcoming in grants from the Directors General of Inland Navigation for extending the canal to the Liffey, and on 10 April 1809, William Gregory, the secretary to the directors, noted that: 'The extension of the Royal Canal to Coolnahay (six miles beyond Mullingar) with the harbour and aqueduct near Dublin, and the docks and communication with the tide water in the Liffey, are finished.'

It does not appear that any provision was made for a bridge across the start of the Broadstone Branch. Horse-drawn boats going to or from the docks at the Liffey would have had to either travel at least half-way down the Broadstone Branch or, alternatively, use the north bank of the canal as the towpath between the Cross Guns and Binns bridges. In September 1933, the Lord Mayor of Dublin, Alderman Alfie Byrne, opened a 'boon for city children [a] tasteful park that had now been laid out by the Corporation on what was some six years ago a canal waterway'. It included an enclosed area for children where they could play in safety. The construction of the park had provided work during a period of high unemployment, and the scheme preserved the line of the branch canal.

The Blessington Street Basin, completed in 1814, supplied the north side of Dublin with water drawn from the Royal Canal, and supplied the Bow Street Distillery (originally owned by Haig's and later by Jameson's) until well into the twentieth century.

In 1845, the Midland Great Western Railway Company (MGWR) bought the Royal Canal. The railway laid its lines alongside the canal most of the way to Mullingar, enabling it to avoid having to conduct lengthy negotiations with individual landowners about crossing their land. The company built a terminal at the Broadstone with a pontoon bridge, which could be moved when boats entered or left the harbour, allowing passengers to reach the station. In 1877, the MGWR filled in the harbour and adapted the drained Foster Aqueduct to provide access to the station. The canal boatmen were provided with a winding hole and a wharf on the east side of the former aqueduct and no longer used the Broadstone Branch.

In the nineteenth century, Broadstone was one of the best known areas of Dublin, although few people even know where it is today. From 1817 the area was one of the major transport hubs in the city, with an important railway station, the canal harbour linked to the Grand Canal, and three dry docks.

The aqueduct and canal that once linked the site to the Royal Canal have vanished almost without trace. The canal was filled in around 1927 and converted into Blessington Park, and the Phibsborough Library was built on top of it.

In 1993, the Dublin Corporation Parks Department began a major restoration of the Blessington Street Basin as an ornamental lake, removing 6,000 tons (6,096 tonnes) of silt and debris, adding a fountain, enlarging the central island for water fowl and wildlife and extensively replanting it. The basin is now supplied with water from above the eighth lock on the Royal Canal, two miles (3.2 kilometres) away.

Ordnance Survey of Ireland Grid Reference: 735083 735502

Broharris Canal

The Broharris Canal was an isolated canal in County Londonderry, Northern Ireland, with no locks. It was constructed in the 1820s when a cut, some 2 miles (3.2 kilometres) long, on the south shore of Lough Foyle, near Ballykelly, was made towards Limavady. As well as serving as a drainage channel, it was used for navigation by barges carrying goods from the port of Londonderry, along with shellfish, sand and kelp for use as fertiliser from the sand banks along the shore of the lough. Its course, which is now little more than a narrow drainage channel, can be traced from the lough, and some way beyond its crossing by the railway, until it is joined by the road from Carse Farm.

An appeal from the inhabitants of Limavady for a canal the whole way to the town from Lough Foyle was turned down and the cut, known locally as the Broharris Canal, was the nearest they came to having a navigable link with the lough. The Irish engineer John Killaly, who had been trained by William Jessop, was commissioned to make a survey, and he estimated the cost of constructing the canal from Lough Foyle to Limavady at £12,000. Despite optimistic estimates of the volume of traffic likely to use the canal, plans to replace it with a horse-drawn tramway were proposed in the early 1830s. Although the canal plan was abandoned, the tramway was never built.

Ordnance Survey Grid Reference: NV 80854 86492

Coalisland Navigation (Tyrone Canal) and Ducart's Canal

The Tyrone Canal, or Coalisland Navigation as it was known locally, was built to carry coal from the collieries at Coalisland, County Tyrone, that were opened in the early part of the eighteenth century. The canal was slightly less than 4.5 miles (7.2 kilometres) long, had a depth of 5 feet (1.5 metres) and had seven locks, including a staircase lock at Macks Bridge, with a total rise of 51 feet (16 metres). Most of the structures still exist, but of the three lock houses on the canal, only one remains. As most of the channel is still intact, a group called 'The Friends of the Coalisland Canal' was formed in the 1990s and has held regular events, such as canal walks and lectures, to raise awareness of the canal's heritage and to explore the potential benefits to the area of re-opening it. The Friends group became a branch of the Inland Waterways Association of Ireland in May 2003 and regularly holds 'clean up' days to keep the canal clear. The most ambitious event organised by the group was a small boats rally, when a fleet of small boats sailed up the canal for the first time in over fifty years.

The canal was intended to enable coal to be transported to Belfast and Dublin (by way of the River Blackwater, Lough Neagh and the Newry Canal) to power the mills and heat the houses of those wealthy enough to afford it. It was estimated that the coal could be sold in Dublin for 6s a ton, about a third of the existing cost in the two cities. Coalisland would also become an inland port for lighters carrying grain for mills and provisions for the surrounding towns.

The idea for the Coalisland Navigation had first been suggested in 1709 by Thomas Knox, a colliery owner and Member of Parliament for Dungannon, but at that time it attracted no support. In 1727, however, the scheme was revived, by which time some 60,000–70,000 tons (609,628–711,233 tonnes) of coal were being imported to Dublin annually from South Wales, and the idea of supplying it instead from Irish collieries gained support. In that same year, Francis Seymour, the owner of a colliery at Brackaville, near Coalisland, published a pamphlet, entitled *Remarks on a Scheme for supplying Dublin with Coals*, which suggested that a canal could be cut across a bog from Drumglass, where many of the pits were located, to join the River Torrent. However, his suggestion also went no further.

Two years later, however, the Surveyor General Arthur Dobbs gave his support to the Coalisland Navigation, although it was not until 1732 that the recently established Commissioners of Inland Navigation for Ireland authorised work to begin.

The navigation was intended to run parallel with, and to be fed by, the River Torrent. Work finally commenced in 1733 with Acheson Johnston as engineer, but it ran into unexpected engineering difficulties, which were compounded by Johnston's inexperience; in fact, he seems to have been only an engineering amateur. According to an account of his work he 'displayed a complete lack of engineering skill'. In spite of this, he would go on to become the manager of the Newry Canal.

Johnston built a large basin at Coalisland, which was supplied by a feeder from the River Torrent. However, with no weir, the current of the river carried quantities of stones and clay which silted up the basin, the locks and the canal itself. Worse still, of the seven locks, the upper two were built on sand, while the lower three were built in a peat bog. In both locations, the walls of the chambers had to be piled and the floors paved, neither of which were done satisfactorily. The lower lengths of the canal were also very close to the river, which flooded it at high water and drained it when it was low.

Thomas Omer suggested a canal could be constructed from Coalisland to the pits, music, no doubt, to Francis Seymour's ears. It would be 3 miles (4.8 kilometres) long and would rise through sixteen locks, with an estimated cost of £15,668. His plans were approved in 1761 but, as the canal was to be capable of handling 100-ton boats, it was wildly ambitious. Omer handed over the project to Christopher Myers in June 1762, without any of the work begun. Myers did build half a mile (0.8 kilometre) of the canal and part of a lock measuring 125 feet by 22 feet (38 metres by 6.7 metres). He then reported to Parliament on the probable cost of the scheme, and recommended that the size of the locks should be reduced to take boats measuring 60 feet by 12 feet (18 metres by 3.7 metres).

Parliament granted £5,000 towards the construction, but wanted a second opinion on the engineering, which was where French-Italian Daviso de Arcort, called Davis Ducart by local people who had probably never previously heard an Italian accent, came in, suggesting the use of two-level lengths of canal, largely in tunnels. Coal would be carried in boxes that could be lowered down vertical shafts into boats on the canal below.

By November 1767, some short lengths of the open-air sections of Ducart's Canal had been built, together with part of an aqueduct over the River Torrent, at a cost of £3,839. Myers went back to the Dublin Parliament for another £14,457 to finish the project, but further discussion followed, the price increased to £26,802, and the tunnels were abandoned in favour of inclined planes called 'dry hurries' or 'dry wherries' by local people. Ducart planned that the inclined planes would have rollers fitted and that the boats would be hauled up using power from a waterwheel. However, when the engineers William Jessop and John Smeaton were consulted, they advised him to counterbalance the boats instead. Even then, the planes could not be made to work effectively, and Ducart replaced the rollers with cradles to carry the boats, running on conventional rails.

Ducart's Canal finally opened in 1777. 1787 saw the closure of Ducart's extension, when it was conceded that the planes were too steep for the cradles to work effectively, and there were also serious problems with water leakage on the upper section. Central control of the inland waterways was abolished by the Dublin Parliament that year. Responsibility for the navigation was handed to trustees, who included the colliery owners. In 1788 there were calls for an adequate water supply for the Drumglass to Coalisland extension, or the provision of a waggon way.

The same year, a Bill was presented in Parliament to encourage the growth of the linen industry by the provision of an efficient means of transporting coal from Drumglass. The linen bleachers imported coal from England, and the feeling ran high that they should use Irish coal. Because they could not get it easily from Tyrone, they called for a complete overhaul of the navigation.

The volume of coal carried on the navigation, however, was lower than predicted, hampered by the lack of a proper link from Coalisland to the collieries, although trade in flaxseed, grain, rock salt, timber, fish and hardware gradually increased. But the condition of the navigation deteriorated, and by 1801 it was almost derelict. Control of the inland waterways was returned that year to the government, and the Directors General of Inland Waterways sent Henry Walker, an engineer, to inspect the navigation. He began work on repairing it, but was dismissed as similar work he had carried out on the Newry Canal was alleged to be defective.

Walker was replaced by John Brownrigg who, from the 1780s, had worked for the Grand Canal Company, where he supervised the laying out of the Barrow branch of that canal. He reported that much of the navigation was in a dangerous condition. The Directors General sought a third opinion from the engineer Daniel Monks, who confirmed Brownrigg's opinion. As a result, between 1801 and 1812, more than £20,000 was spent on scouring the basin at Coalisland, building wharfs, stores and boundary walls around it, rebuilding the locks, dredging the whole waterway to a depth of 4.5 feet (1.4 metres), repairing the lock houses, puddling the lower lengths where the navigation passed through peat, and upgrading the towpaths.

Following all this work, which had put the waterway in the best condition it had ever been, traffic carrying a wide variety of commodities steadily increased, but coal traffic did not. This was a direct consequence of the passing of the Act of Union in 1800, which increased the free movement of trade between Ireland, England and Wales. Dublin had since imported all of its coal from England, which, though slightly more expensive, was of much higher quality.

Meanwhile, with the decline of the collieries at Drumglass, Coalisland had developed as an industrial centre, importing raw materials along the navigation and exporting tiles, bricks, earthenware, shovels, vitriol and sulphur, all of which were manufactured locally. However, the distribution of these goods was hampered by the narrower locks and shallower depth of the Ulster Canal, which connected the River Blackwater to Lough Erne.

Responsibility for the navigation changed yet again in 1831, passing to the Board of Public Works. Traffic continued to increase, from 8,200 tons (8331.5 tonnes) in 1837 to 18,888 tons (19,191 tonnes) in 1866. After that, however, railway competition began to have a growing impact on traffic. Although the navigation continued to be maintained in good order, the costs exceeded receipts from tolls and rents. In order to avoid having to subsidise the navigation, the government looked for a buyer and finally succeeded in selling it to the Lagan Canal Company in 1888.

The new owners increased the navigable depth to 5 feet (1.5 metres), enabling barges of 80 tons (81.2 tonnes) burden to reach Coalisland, as a result of which the export from the town of sand, bricks, agricultural products, pottery, fireclay goods and timber to Belfast increased significantly, as did imports of grain, coal, hardware, foreign timber and provisions. Between 1890 and 1900, total traffic increased from 18,000 tons (18,288.8 tonnes) to 36,000 tons (36,577.6 tonnes), and an overall loss of £89 per annum became a profit of £355 by the end of the decade.

The Lagan Canal Company planned to increase the depth of the navigation to 5.5 feet (1.7 metres) to enable vessels to work through from its canal, but the start of the First

World War in 1914 meant that the economy had to be devoted to the war effort. Once the war was over, both the Lagan Canal and the Coalisland Navigation faced growing competition from more flexible road transport. From 1 July 1917, the Coalisland Navigation, together with the Lagan Navigation and the Ulster Canal, came under the direct control of the British Government, but when they were returned to private ownership, the navigation shared an improvement grant of £19,000 with the Lagan Canal.

Despite competition from both railways and road, traffic in building sand, grain and coal continued to increase through the 1920s, reaching 57,000 tons (57,914.6 tonnes) in 1931, when tolls raised £1,634 and the company made a profit of £650. After that, however, traffic declined rapidly, and in 1939 the company made less than £50.

During the Second World War, the navigation saw very little traffic, and it ceased altogether in 1946. The navigation was officially abandoned in April 1954, when control of it passed to the Ministry of Commerce, and it became nothing more than a drainage ditch.

The basin at Coalisland was drained in 1961 and now lies underneath the car park of the Coalisland Heritage Centre. One lighter, *Eliza*, was buried where it lay. The rest of the waterway is owned by the local council and the Department of Agriculture. The former Coalisland Navigation has left a very visible legacy to the town in the form of a five-storey former corn mill, which stood by the canal basin and dominates the town.

One local man's earliest memory of the navigation was of:

… standing on the footpath in front of our house and shop at Harbour View and catching the grains of Indian corn as they bounced from the high-sided carts carrying the grain from the lighters in the Basin to John Stevenson & Co's mill on the Dungannon Road. The carts were painted with orange lead and pulled by large Clydesdale horses, which were so well used to the routine of drawing from the lighters that they could have done without the carters who accompanied them. The grain was scooped into a barrel and lifted from the cargo hold of the lighters by a tripod [sheerlegs] and then tipped into the high carts on the quayside. It all looked well-regulated and orderly as each carter strode along at the horse's head, with his arm through the reins as he filled his pipe with 'Warhorse' tobacco and lit up.

Stewart's Mill didn't need horses and carts: their mill was built right on the edge of the Canal and the grain was lifted by elevator into the mill from the lighters drawn alongside. Further down Lineside, imported coal was lifted from lighters for Kilpatrick's Coal Yard nearby, a reversal of the role for which the canal was built in 1733 (to bring Coalisland coal to Dublin).

Across the Basin on Main Street, sand was dropped down a chute into the lighter below for transport to Belfast, while at the top of Washingbay Road the men from Ulster Fireclays carefully loaded their clay sewer pipes and bends of all shapes and sizes into the hold of a lighter moored opposite the works.

How times have changed. Since the fifties Harbour View has become "The Square" and the Basin has been drained and filled in. About half of it is used as a car park, a toilet block and a roadway, which links Lineside with Main Street. The remainder is a grassy area and, from the second lock, the Canal lies stagnant to the Blackwater River, 4½ miles away.

Ordnance Survey Grid Reference: NV 95933 26243

Cong Canal

The Cong Canal in County Mayo, known, appropriately, as the 'Dry Canal', was a failure due entirely to its inability to hold water. Now used as a drainage channel only, the water level can vary between 6 inches (15.2 centimetres) and 12 feet (3.6 metres) depending on the time of year; in summer it is dry and in winter it is full. Some built features of the canal remain.

A walk along the dry canal brings reminders of the Great Famine of 1845–48. The canal was conceived as a famine relief scheme that never fulfilled its promise in any sense of the word. Because of the porous limestone underlying the area, the water disappeared into the ground like it was running down the plug hole of a bath.

The construction of the canal was paid for by Benjamin Guinness, whose intention was to join Loughs Corrib and Mask and create a safe transport link from Sligo to Galway, avoiding the need to pass along the stormy west coast of Ireland. Work began in 1848 and continued until 1854, when the decision was finally taken to abandon the work, except for that necessary to allow the channel to serve as a drain.

Today, north of Cong at Creggaree, the canal is diverted into an adjacent stream and the bed remains dry from there to Lough Corrib. The section south of Cong was sold to Lord Ardilaun, becoming part of the Ashford Castle Estate, and one of the locks was converted into a boathouse.

The canal was intended to join Lough Mask to Lough Corrib, both for drainage and navigation. The Commissioners of Public Works asked their engineer John McMahon in 1844 to assess the proposal. McMahon reported favourably in 1846, and Samuel Roberts was appointed superintending engineer in 1848, beginning work immediately. In April 1854, however, the commissioners decided that the navigation aspect of the project should be discontinued, although drainage work was to continue, as it was recognised that it would enable the level of Lough Mask to be regulated to prevent flooding.

The porous limestone was known to the engineers, and they actually used it to their advantage during the construction of the canal. During the winter floods, they worked on the canal near Cong and, in the summer, they used the disappearance of the water down swallow-holes to work on the upper reaches of the canal. In his report for 1852, Samuel Roberts wrote:

> During the early part of the season while the flood waters of Lough Mask were being discharged through the upper level of the canal and overflow course, the excavation for the lower levels and the several works of masonry in connexion with them, were proceeded with and have been continued throughout the entire season … When the flood waters of Lough Mask had subsided to the level of 68 feet [20.7 metres] above datum, the head dam was made good across the channel at the debouch from Lough Mask, and the discharge of

that lake confined to that effected through the cavernous passages; the excavation necessary for the completion of the upper reach of the canal was then recommenced, and has been carried on throughout the season with the greatest vigour, the excavated material has been removed by waggons to form the trackway embankments through the inner basin of Lough Mask … The channel runs through cavernous limestone rock for its entire length of 4 miles and 40 yards [7 kilometres], and the depth of cutting varies from 8 to 20 feet [2.4 to 6.1 metres].

The main structures that were built included three locks, each measuring 94 feet by 16 feet (28.6 metres by 4.8 metres), two stone bridges, one at Cong and one at Drumsheel, and an aqueduct between Locks 2 and 3. There was an 'overflow course', as well as a set of regulating sluices, towards the Lough Mask end.

The channel between the quay and Lough Corrib had to be deepened. Roberts's report for 1853 said:

The spoil taken from the excavation of the channel has been formed into a bank on the east side, 40 feet (12.1 metres) in breadth of which has been reserved for a trackway and landing place for boats. On the west side of the channel a guard wall has been built for the purpose of diverting the current of the river in a direction favourable for the navigation of boats entering the lock.

The trackway had a boundary wall 4 feet (1.2 metres) high. The channel at the quay was improved in 1856 and again the following year; an 80-foot (24.3-metre) wharf was built immediately below the lock entrance in 1856, with a road along the trackway to Cong.

The second lock was known as Cong Lock and it appears that it was in the village. Lock 3 was only a few yards from Lock 2. The road that entered Cong from the east was realigned to run alongside the canal, crossing it below Cong Lock, after a right-angle bend, on a bridge with a span of 16 feet (4.8 metres) and a roadway 24 feet (7.3 metres) wide, which meant that there was no towpath under the bridge.

Creggaree Lock, and the aqueduct, lie between Locks 2 and 3. The three locks are grouped together at the Cong end and the double lock on the Ardnacrusha headrace canal is at its lower end. In 1844 the owners of four mills at Cong expressed interest in having their water power improved, so it may be that the grouping of the locks maintained the head of water and thus maximised the power available to the mills.

Drumsheel Bridge was described as having a 26-foot (7.9-metre) span and a roadway 12 feet (3.6 metres) wide, so, unlike the bridge in Cong, it was wide enough for a towpath.

Stone from Drumsheel was used elsewhere on the canal. Roberts' report for 1852 said:

The excavation of the half width of the canal at Drumsheel, which remained since preceding years, has been carried on for the purpose of procuring the necessary material for the erection of the locks and bridges – the stone at this place being the best suited for the work.

The original plan was that, for most of its length, the canal would form the main drain from Lough Mask. From just above Drumsheel and the start of the locks, though, the canal would carry enough water for navigation. Any excess water would be drained by a weir into an overflow channel, and would be carried to Cong to power the watermills.

Ordnance Survey of Ireland Grid Reference: 514828 755297

Eglinton Canal

The Irish Parliament began to consider inland navigation, using canals to supplement the navigable rivers, as early as 1703. As was the case throughout the British Isles, road conditions were usually appalling, making the transport of goods and commodities around the country extremely difficult. In England, the impetus for building canals came from private enterprise. In contrast, however, those with money in Ireland kept their hands in their pockets and waited for the government to react to public demands for a more efficient means of transport. Obligingly, the Dublin Parliament first passed the necessary legislation for the construction of inland navigation schemes in 1715.

The idea of a canal to connect Lough Corrib with Galway Bay was first proposed in 1822 by the engineer Alexander Nimmo. However, there was not enough support for the scheme and the idea was abandoned before any work was done. Work, funded by the Commissioners of Public Works, eventually started on the construction of the Eglinton Canal (named after the Earl of Eglinton, the Lord Lieutenant) on 8 March 1848. For the people of Galway City it was particularly welcome, as the work would provide much needed employment for those suffering the effects of the Great Famine.

When the work was under way, the District Engineer, Mr S. U. Roberts, wrote:

> The boundary walls have been built, and the excavation of the canal proceeded to a considerable extent … The total quantity of excavation in the canal cut, which is chiefly rock, is 55,356 cubic yards (42,323 cubic metres), of which 23,700 cubic yards (18120 cubic metres) have been excavated and removed.

Work also began, using ashlar stone, on the construction of the tidal basin, which became known as the Claddagh Basin. Dams also had to be constructed so that the work could be carried out. The basin would connect the canal to the sea with lock gates. The canal basin at the Claddagh, which has well-built limestone walls and steps, still survives. An integral part of Galway's industrial and transport heritage, it was formerly an important part of the infrastructure for the city's waterborne trade.

By 1849, considerable progress had been made. Most of the canal had been excavated and work was also carried out in excavating the River Corrib itself at Newcastle, Menlo, and the twelfth-century 'Friars Cut' to a depth necessary for the navigation of bigger vessels. Lock gates and swivel bridges had been built, and work on the lock-keepers' houses, at Dominick Street in the city close to Ball's Bridge and at Cong, was also in hand. The first lock-keeper in Galway was John Keogh, and his salary was £45 12s 6d a year.

The canal was only about 0.75 miles (1.2 kilometres) long, but it was planned that it would be 8 feet (2.4 metres) deep, and 40 feet (12.1 metres) wide at the bottom, throughout its length and with a total rise of 14 feet (4.2 metres). It followed a semi-circular course around the west of Galway city, bypassing the main line of the river and many other watercourses. Other improvements were made to Lough Corrib at the same time. The original estimated cost was £38,000, including £11,000 to improve the flow to watermills in the city by the construction of new 'tail-races'. In the mid-nineteenth century there were twenty-seven watermills in Galway, eight of which seemed to have been powered by water from the canal, fed through the Convent River.

The canal had two locks, the Parkavera Lock, some two-thirds of the way from the upper end, and the sea lock that connected the Claddagh Basin to the bay. The locks were 130 feet by 20 feet 6 inches (39.6 metres by 6.2 metres). There were also five hand-operated swing-bridges.

There was great excitement throughout Galway in anticipation of the opening of the canal, which had been built in record time. On 15 September 1851, after all the debris of construction had been cleared from the canal, it was opened to allow the water to flow in.

Even the *Illustrated London News* sent a reporter and an illustrator to record the historic event. The report said:

> Their Excellencies, the Earl and Countess of Eglinton, after receiving the address of the Harbour Board, and enjoying a short sail in Galway Bay, returned ashore at the Pier and were driven to a small steamer … Having gone on board the *O'Connell* amidst the sounds of music and cheers of the people, deputations and addresses were presented … the steamer entered the dock for the first time amidst the cheers of thousands. Here a long procession was formed of the fishing hookers and other boats belonging to the Claddagh, with those of the Board of Works and private parties in the town, headed by the *O'Connell* and took its way towards the first gate, which closed, to raise the Viceregal party's boat to the level of Lough Corrib. Here the view was obtained of the Lord Lieutenant and his Lady. The steamer made its way up to the Castle, from which a royal salute was fired and afterwards put about for the downwards progress.

In the following years, traffic through the canal was steady, with some 6,800 tons (6,906 tonnes) of goods being carried in 1857. By 1881, however, there was growing concern among the people of the town about the dangers of the canal. The *Galway Express* reported that more than eighty people had drowned because the canal had been left unprotected. In response, the council decided to erect railings along the length of the canal. This was welcomed in a report on the matter which said:

> It is gratifying to see that at last steps have been taken towards remedying an evil which is long-outstanding in Galway, namely, the unprotected state of the canal, where so many persons have met with untimely graves. A more dangerous watercourse could hardly be constructed than that which runs along the western portion of the town, and which until lately was entirely unprotected, though its banks on each side are in fact public thoroughfares for pedestrians … now, to the credit of those who are the trustees of this useful, though undeniably most dangerous Canal, a railing is being erected … which will

effectually, it is to be hoped, prevent a further repetition of those melancholy occurrences which almost weekly take place in Galway.

Among the many vessels to use the canal were the *Enterprise, Fr. Daly, Lioness, Lady Eglinton, Saint Patrick, Fairy Queen, Countess Cadagon* and the *Corrib Advance*, which carried passengers, coal, meal, flour, bran, grain, manure and timber. 300 fishing boats operated out of the Claddagh basin.

The canal was useful, not only for the movement of vessels, but also in draining and regulating the surplus waters of Lough Corrib, and in enabling vessels to enter the lough from the sea. In addition it helped to lower the winter flood levels of the lough. A deep tail-race was also built from the distillery at Newcastle to discharge into the river.

A 1904 financial report put the total revenue for the canal at £992, which represented 3,194 tons (3,245.2 tonnes) of goods. It was, however, a marked decline from the high point of 1857. The income from tolls continued to decline, from £35 in 1905, to £10 in 1915 and just £1 in 1916.

Over the following decades, neglect and a general lack of interest saw business on the canal fall dramatically. To make matters worse, the gates and bridges required maintenance to keep them in safe working order. The canal's trustees had to face up to the fact that they could no longer afford to keep the canal open.

The five swivel bridges across the canal were replaced with fixed bridges, meaning that goods traffic could no longer pass through. The last vessel to use the canal was the Guinness's 90-foot yacht *Amo II*, a former First World War mine-layer, and the Eglinton Canal finally closed in 1954. Today, people enjoy leisurely walks along the now tranquil waterway close to the heart of the old city.

Ordnance Survey of Ireland Grid Reference: 529570 724752

Finnery River Navigation

The Barrow Navigation runs from Athy, in County Kildare, down to a sea-lock at St Mullins. From there, vessels can go downriver on the tidal section of the Barrow, and then up the River Nore to Inistioge or down through New Ross to join the River Suir near Cheekpoint. Along with the Rivers Suir and Barrow, the River Nore is one of the group of rivers known as the 'Three Sisters'.

The town of Athy is dominated by the well-preserved sixteenth-century turreted tower of White's Castle, which stands by the bridge over the River Barrow in the heart of the town. The Grand Canal linking the town with Dublin was extended to Athy in 1791 but, with the opening of a railway on 4 August 1846, the canal fell into disuse.

The Barrow Navigation links Athy to the main line of the Grand Canal at Lowtown. Although the town is the head of the navigation, there was at one time traffic further upstream, including cargoes of peat from the bogs around the Finnery River. Brian Goggin notes that the Ordnance Survey map of 1900 even shows a lock in the mill weir above the main bridge in Athy.

The Literary Panorama and National Register in June 1815 reviewed the third and fourth reports of the commissioners appointed to enquire into the nature and extent of the bogs in Ireland, and to assess the practicability of draining and cultivating them. In an appendix to the third report, Richard Brassington's survey of a 'District on the River Barrow, in Kildare and King's County' said:

> The processes recommended for improving these bogs are various: drainage is the leading feature of them all. The idea of converting the drains necessary for that purpose, in some places, into navigable canals, is happy. The system of drains in others is judicious.

The importance of bogs historically had increased for two reasons: the disappearance of woodland, and the resulting scarcity of wood as a domestic fuel, and the pressure exerted by a growing population. Peat was not considered to be an inferior fuel. In medieval times it was burnt in monasteries and manor houses as well as in the cottages of the common people.

On the Ordnance Survey map surveyed in 1837 and published in 1839, the Finnery is shown splitting in two just before it joins the River Barrow. One of these branches is straight and gives the appearance of being artificial. Upstream of Cloney Bridge, it bends sharply and runs directly north, parallel with the road. It passes under another bridge on a local road between Boherbaun and Cloneybeg, called the Black Bridge on the 1908/09 OS map. The bridge does not appear to have be intended to be navigable. Shortly after the bridge, and still following the river upstream, the Finnery turns north–north-east and just above the bend is what could have been a winding-hole. It then heads north-east in the direction of what used to be bogs, which have been drained and replaced by a plantation of trees. The river then passes into a complex of what appear to be small drains and then re-emerges on the far side of the bog.

Brian Goggin's belief is that at one time the Finnery was both a drainage channel and a navigation, probably carrying peat from the bog to Athy.

Ordnance Survey of Ireland Grid Reference: 665556 702705

Grand Canal Docks (Abandoned Line)

The original main line of the Grand Canal terminated at the Grand Canal Harbour near St James's Gate in Dublin City. Most of the route of the canal today is used by the red line of LUAS, the Dublin area tram or light rail system.

The harbour was where the Dublin Gas Company originally had its main wharfs. This was where colliers from Liverpool delivered the large quantities of coal that were required, both for industrial and domestic use. In those days, before the discovery of natural gas, all gas used by both industry and domestic premises was coal gas.

The gas company's location meant that there was also plenty of water, which was necessary in the process of separating gas and coke from coal. Later, when the company's coal requirements increased, it transferred to wharfs at Sir John Rogerson's Quay, which could take larger ships. By the 1880s, the gasworks dominated the landscape from the Barrow Street gasometer to Rogerson's Quay.

As the demand for coal from across the Irish Sea grew, the Grand Canal Docks became an increasingly convenient location for all the importers of coal. It was not only useful for customers in South Dublin, but also for supplying users of coal all along the Grand Canal and the Barrow Navigation. In 1918, the large dry dock was filled in and the ground leased for a coal yard. Two graving docks were also filled in, and the land was taken over by the gas company.

During the First World War, it became ever more difficult to maintain the vital supplies of coal from Britain. In 1916, the Alliance & Dublin Gas Company acquired its own ship, the 440-ton *Ard Rí*, and then a second vessel, the 400-ton *Braedale*. The *Ard Rí* still docked on Sir John Rogerson's Quay but the slightly smaller *Braedale* was capable of passing through the Grand Canal Locks to moor alongside the company's depot. The company had a new ship built in 1920, the 460-ton *Glenageary*, which was designed to use the 492-foot (150-metre) long Camden Lock, and the Dublin Gas Company was again able to use the Grand Canal Docks. From 1919 to 1968, it had its coal depot on the site of what are now Gallery Quay and Grand Canal Square.

The *Glenageary* became the flagship of a fleet of four steamers, the others being the *Glencullen*, the *Glencree* and, from 1946, the *Glenbride*. With a crew of eleven, each of these steamers carried coal from Liverpool to the Grand Canal Docks, a voyage that took between sixteen and eighteen hours.

The Grand Canal's line to its harbour was also used by the Guinness brewery, both for transport and for supplying water to the brewery and the houses on the south side of Dublin. Today, boats entering Dublin on the Grand Canal from the west pass the filter beds above the eighth lock. There is also a second set of filter beds at the fifth lock. The beds still supply water to Guinness, although now the water is only used for washing in the brewery. In the past, Guinness was made not with water from the River Liffey, as many people believed, but with water from the Grand Canal or the Royal Canal. Other breweries and distilleries also took their water from the canals.

The Guinness brewery is right next to the Grand Canal, meaning malt can be brought in and stout can be taken out. The brewery is close to the City Basin, and originally the company had the right to draw water from a culvert that had carried water from the River Poddle to the basin. From 1777, the canal itself fed the basin. When Guinness expanded into the area north of James's Street, as far as Victoria Quay, it was able to build its own quays on the Liffey, where its barges could carry the barrels downriver to its own ships.

The section of the canal that narrowed as it passed the south end of the basin, and curved north to the harbour, was known as 'The Gut'. In 1974, a journalist from the *Irish Times* asked local residents whether they thought that the, by then derelict, line of the canal should be filled in. Almost everyone interviewed was in favour of it being filled in, mainly because of the danger that children might drown, but also because it encouraged rats.

The curved part of the harbour became a graveyard for wooden boats that were no longer needed, and sugar was also unloaded there. This harbour was infilled in 1960. The middle harbour was used for loading and unloading, mostly at the western end, because the eastern end was open to the road and pilfering would have been inevitable. The western quay was used for loading full barrels of stout and unloading empty ones. This harbour was also infilled in 1960. The southern harbour had a crane with which stout barrels were loaded and unloaded, but boats rarely moored at its northern end where sugar was unloaded and stored. Spare boats were moored in this section and the three dry docks allowed boats to be serviced, repaired and painted.

The last working cargo barge passed through the original Grand Canal in 1960.

Ordnance Survey of Ireland Grid Reference: 713783 733934

The Grand Canal in Dublin. Image used by permission of William Murphy.

John's Canal, Castleconnell

The Stein & Browne distillery at Thomondgate, Limerick, was in business by 1809. It was very productive, distilling 231,790 gallons of spirits in 1822, which was about one-twentieth of the total production of all Irish distilleries – of which there were many – that year. Five years later, it produced about one-sixteenth of the total output.

Most historians describe the Steins as a great whisky dynasty, probably one of the greatest understatements you will ever read. The Steins revolutionised both the Scottish and Irish whisky industries, but, without a doubt, the members of the family were also some of the greatest Scottish industrialists of their time. The distilleries founded by the Steins were the largest manufacturing undertaking of any kind to emerge during the first decade of the Industrial Revolution in Scotland and Ireland. Although the family name Stein sounds Jewish, it derives from the town of Stein on the Isle of Skye.

In Scotland, the Steins had distilled whisky on a scale never seen before. They were ready to embrace any new techniques that would benefit distilling. Two of their greatest achievements were that they founded the export market for Scotch whisky and the invention of the 'continuous still'.

Distilling on such a large scale created the problem of how to keep the plants supplied with the raw materials. This in turn led to great changes to farming in the surrounding area. The whole infrastructure of central Scotland had to be improved and the Steins confidently commissioned large engineering projects, including the construction of roads, canals and tramways. The Steins were not short of the necessary drive and determination, or even ruthlessness, needed to drive forward these enterprises.

John Browne (sometimes spelt Brown), Stein's partner in the Limerick distillery and who also had a share in Andrew Stein's Marlfield Distillery in Clonmel, gave evidence to several Parliamentary Committees during the 1820s and 1830s. In 1837, answering questions from the Select Committee on the state of the poor in Ireland, he said that he employed a great many men as labourers in cutting turf, at a bog near Castleconnell, on the River Shannon side of the road between Castleconnell and Montpelier. Browne's management of the bog represented the best practice at the time.

The Stein & Browne distillery installed what was probably Limerick's first steam engine in 1822. It is not known exactly when the partners bought the bog, but the steam engine must have been running on turf by 1830 at the latest, and possibly did so from its installation.

It seems that John Browne retired from the distillery, or died, some time between 1840 and 1846. During the Great Famine, the firm's principals were two Messrs Stein and a Mr Walnutt. The latter was charged with substituting inferior corn of his own for the firm's corn when supplying the Relief Committee of Broadford and O'Callaghan's Mills.

On 20 September 1841, the lease on the bog passed to James Macnab, one of three Scottish brothers who went to Ireland. Two of them, including James, had worked with their father at a distillery owned by Stein in Alloa, and they came to Ireland to work at the Limerick distillery. James was put in charge of the bog round about 1823. When he died in 1865, his son Alexander Macnab took over. In the 1870s, he became an authority on the preparation of turf as fuel. He emigrated to the United States in 1885. The bog was sold in 1897.

The Ordnance Survey map, 1840, shows a canal (which was possibly called John's Canal, although no name is given) leading into the bog with two branches, which by the time the 1893–99 map was surveyed had become a more extensive system of channels, serving the dual function of draining the land and transporting the turf.

Writing about it in the *Irish Times* in 1873, Mr J. McC Meadows ME, a member of the Irish Peat Fuel Committee, cited the Macnab operation of the bog as an example:

There the cutting and saving of turf as a marketable commodity has been followed for many years. Fifty years ago this land had more the character of a morass than of a firm bog, but this circumstance did not discourage enterprise. A substantial pier for boats was built; from the Shannon a canal was cut into the morass, with branches for drainage and transport. Some three hundred persons were soon employed every season in the cutting and saving of turf.

In 1842 the property passed into the hands of Mr James Macnab, who from the first conducted its management, and from that time to the present the raising of turf for sale to the public, coupled with the reclamation of the bog, has been followed as a "special business". The cutting commences in March, and is continued until August; and five thousand tons are made and sold annually.

Mr A. A. Macnab, the present proprietor, states that wet weather does not prevent the safe harvesting of the turf, and during the dripping summer and autumn of 1872 he cut and saved in good condition five thousand tons, all of which was disposed of to the public a month or two before the end of the year [...] The present state of this tract of country, which, within the memory of many now living, was a waste, is instructive and encouraging.

It is said locally that there was a lock gate at the entrance to the bog canal which joins the Shannon just upstream of, but on the opposite side of the river from, the upstream end of the Errina–Plassey Canal. This would have made the carriage of turf by water to Limerick very easy.

It is not clear whether the turf went from the canal harbour in Limerick to the distillery by road or by water. Turf was usually sold from the left side of the harbour, opposite the Lock Mills, so the distillery's turf could easily have been unloaded there to be carted to the distillery. Alternatively, the boats could have been worked up the Abbey River, then down the main stream to Browne's Quay at the distillery, just above Thomond Bridge.

The second report of the Railway Commissioners, in 1838, lists turf boats belonging to the Limerick distillery as one of the four main types of boats using the Limerick Navigation from Limerick to Killaloe. Little is known about the types of boats in use or how they were propelled, however it is unlikely that they carried more than 16 tons

(16.2 tonnes), and possibly only 5–10 tons (5.08–10.1 tonnes). There is, however, a painting in the Limerick City Gallery of two boats which seem to be turf boats – one empty and the other full – being propelled by men with poles. They are in the river off Browne's Quay, above Thomond Bridge, although the distillery is out of sight to the right.

It appears that the distillery may have ceased using turf, perhaps after Browne left for America, and that Macnab then began selling turf to the public. Alexander Macnab wrote about the advantages of water transport, so it is quite likely that he would have continued sending the turf by water into Limerick.

Ordnance Survey of Ireland Grid Reference: 566359 662449

Lismore Canal

In 1792, the 5th Duke of Devonshire's resident agent, Henry Bowman, undertook the task of revitalising the Lismore Castle Estate and improving the trade of Lismore through the construction of a canal linking it with the navigable part of the Blackwater River. A creamery and butter factory that still survives was also built on the canal in 1793. Known as the Buttermill, it is a four-storey building of sandstone. In plan it is U-shaped, with one block fronting on to the canal basin and two other blocks facing north towards the main road, with an archway through the centre for access to the edge of the basin.

The Lismore Canal, dug in 1793 to link the town of Lismore with navigable reaches of the River Blackwater further downstream, was in occasional use until as late as 1922, when the railways were severely disrupted by IRA activity, and is still partially accessible to small boats.

The Lismore Estate was involved in a number of transport improvements, including the building of the canal for trade in timber and coal and to transport livestock. Cattle, sheep, pigs and beet made up most of the general cargo together with bacon, and salmon caught in the Blackwater. Another important cargo for a period was the Derbyshire gritstone which was used to face the castle, the railway station and a number of other buildings in the town.

In 1814, William Cavendish, the 7th Duke, built a short canal from Cappoquin, at the head of the tidal section of the river, to Lismore. There were therefore three waterways in the area: 16 miles (25.7 kilometres) of the Blackwater from Youghal to just above Cappoquin; 7 miles (11.2 kilometres) of its navigable tributary the River Bride; and the Lismore Canal. None of these waterways are connected to other Irish inland waterways, other than by the sea at Youghal. Navigation on both the Blackwater and Bride rivers is tidal.

The Lismore Canal, which had a single lock, went from a point upstream of the Kitchenhole to Lismore. The canal still runs next to the road for much of its route.

When the railways made many canals and navigations obsolete, the Duke of Devonshire was the main shareholder and chairman of the 43-mile (69.2 kilometres) Waterford, Dungarvan & Lismore Railway in 1872, which became known locally as the 'Duke's Line'. In 1923 the line was amalgamated with other railways to form the Great Southern Railway.

Ordnance Survey of Ireland Grid Reference: 604825 598974

Lismore Canal. Image used by permission of Mark Morton.

Lynch's Canal

When the construction of the Eglinton Canal began in 1848, the idea of linking the River Corrib with the sea by canal was not a new one. As early as 1498, Andrew Lynch, who was the fourteenth Mayor of Galway, attempted to connect the Corrib with Lough Atalia, which today lies between the suburb of Renmore and the centre of Galway city.

Lynch was a member of one of the 'Tribes of Galway', fourteen merchant families who dominated the political, commercial, and social life of the city of Galway between the mid-thirteenth and late nineteenth centuries. In the late fifteenth and early sixteenth centuries, Galway was engaged in a trade war with Limerick. The merchants of Limerick were licensed by King Edward IV to arrest any men of Galway found in Limerick and to impound their goods.

Some people believe that Lynch's Canal was to be constructed from the Sandy River through Galway town to meet the sea. However, although it is thought that work on the canal began, it was never completed and inevitably the Mayor's project came to be known as 'Lynch's Folly'. Its route, and any evidence of its construction, lie beneath that part of Galway city between Lough Atalia and the River Corrib.

Ordnance Survey of Ireland Grid Reference: 31386 25829

Mountmellick Line (Grand Canal)

For more than a hundred years, the Mountmellick Canal played an active part in the lives of the people of Monasterevin, in County Kildare, and Portarlington and Mountmellick in County Laois. In its first years, the canal carried passengers and brought materials and coal to Kildare and Laois. Later its main cargoes were flour, malt and porter.

During the Industrial Revolution, steam engines enabled mills, mines and factories to produce goods of all kinds at a rate that could never have been imagined a hundred years before. However, in areas with no navigable rivers, the carriage of both raw materials and finished products was a problem. Man-made canals were the perfect solution, especially when it came to transporting coal to fuel factories and agricultural produce to feed the growing population of towns and cities.

Governments, landowners and industrialists – among them Quakers, both in England and Ireland, who were in the forefront of many new industries – all invested money to build canals. Many speculators also invested in the hope of making a killing from their share of profits from goods and passengers and the means of carrying them. In Ireland, especially, it was hoped that if the government helped to create the man-made waterways to transport manufactured goods, industry would grow.

Like the rest of Ireland, Laois was mainly agricultural, but it made sense to open a branch of the Grand Canal to Mountmellick because, since the middle of the eighteenth century, the town, with its large number of Quakers, had been known as the 'Manchester of Ireland'. As a result of Quaker enterprise, by the time canals were being built in Ireland, Mountmellick had become the wealthiest and most industrialised town, and had the greatest population, in County Laois. In 1801, nearly 4,000 people worked in a number of large spinning and textile mills in and around the town. Textiles were also, literally, a cottage industry in the surrounding area and many farmers and their families supplemented their earnings by weaving at home.

Mountmellick's breweries sold beer to people in the town and the rest of the county. The town also had a number of tanneries. A thriving textile mill in the nearby town of Portarlington also expected to benefit from having access to the canal. New Mills, the first integrated woollen mill in the county, was operated by John Millner on a site just west of Mountmellick. In the early 1800s, Anthony Pim also had a maltings and brewery in the town.

In the 1820s, the Irish Parliament agreed to lend the Grand Canal Company money to build a canal from the Athy Branch of the Grand Canal at Monasterevin to Mountmellick. The contractors Henry, Mullins & McMahon, who had been responsible for cutting most of the Grand Canal main line, were hired to construct the new branch, supervised by the engineer Hamilton Killaly, whose father, John, was an engineer for the Grand Canal Company. Beginning in March 1827, the construction of the canal proceeded at an average rate of 27 yards (25 metres) a day. Because the canal followed the floodplain of the River Barrow, it was traversing relatively level ground. Only three locks had to be built, and the terminus at Mountmellick had a large basin and warehouses.

The only large river, the Triogue, was crossed by the solidly built, triple-arched Mountmellick Aqueduct. It was also necessary to construct nine single-arched culverts to cross streams and drains, seven of which are still conveying water nearly 200 years later. Twenty-seven bridges were built along the Mountmellick Branch, carrying roads and railways over the canal, twenty-three of which still stand today.

By July 1829, the canal was ready to receive water but, because of its porous gravel bed, it was another two years before it held water from end to end. It finally opened in March 1831 and, for the next sixteen years, it had a monopoly on the carriage of passengers and goods traffic. Travel by canal barge was much slower than by stagecoach, but much more comfortable: up to twelve people would crowd into a single stagecoach, with a further eight riding on top. On long, dusty journeys, travellers literally rubbed shoulders as the coach lurched over the uneven roads. By contrast, the passenger barge, which was 52 feet by 9 feet 10 inches (16 metres by 3 metres), glided gently along the canal at 4 mph, with its passengers seated on upholstered benches. Up to forty-five passengers could ride in the 'State' (First Class) cabin and, if the weather was pleasant, they could enjoy the passing scenery from a deck on top of the cabin. Another thirty-five passengers sat back-to-back on a central bench and benches along the walls in the 'Common Cabin'.

After 1834, the Grand Canal Company introduced flyboats drawn by teams of four fast horses, which were changed frequently to achieve speeds up to 10 mph, including time spent at locks. There was a schedule of flyboats in the day and ordinary passenger boats by night. One contractor told the Grand Canal Company that he needed sixteen horses to service his stage of 12 miles.

At the Mountmellick end, the canal was topped up by a 2.5-mile (4-kilometre) channel off the Triogue River at Kilnacash. The other feeder was 2 miles (3.2 kilometres) long, originated in boggy ground at Kilbride Wood, and flowed into the canal at Woodbrook Bridge. Today both feeders are largely dry and overgrown.

Coughlan's Lock in Coolnafearagh, County Kildare, was the first of three locks on the branch. The Lock House there is the only one on the canal that is still inhabited. When the Portarlington ring road was built in 1970, the length of the canal that included the second lock was infilled. All that remains of the lock is part of the north-west side wall. No trace of the line exists near the third lock at Tinnakill and the lock has been infilled.

The canal had been very expensive to build, so fares on the passenger barges had to remain high. Although trade on the branch built up gradually from 1831 to 1834, profits were never more than £100 per annum. But for the towns of Mountmellick and Portarlington, the traffic on the canal, as well as bringing in goods, made it possible for businesses to transport their products for sale in Dublin.

When the potato crops began to fail in 1845, Mountmellick was one of the hardest hit towns in Ireland. As food prices rose, people rioted in the streets. In February 1846, the board of the Grand Canal Company agreed to let provisions pass toll-free on the canal to alleviate the distress of the people. By March 1847, 3,300 people were dependent on the

Mountmellick Relief Committee for food. Typhus and cholera spread and, by the end of the famine years, more people in Mountmellick had died from the epidemics than from starvation.

The Quakers Joseph and Margaret Beale played important roles in Mountmellick's history during the 1840s. When the potato crops began to fail, Margaret and her relatives ran a soup kitchen for the poor. Among their other business interests, the Beales owned Monordree (Monordreigh) Mill, a woollen mill that was known for its 40-foot-high (12-metre) waterwheel. When the Great Famine struck, Joseph Beale converted the mill into a flour mill and ground imported Indian corn to feed the people facing starvation.

Eventually, Joseph went bankrupt and emigrated to Australia. Margaret remained behind to manage the mill, which had gone back to producing textiles, for another two years before leaving with the rest of the family to join her husband.

The Famine devastated passenger traffic on the canal, and then the railways dealt the final blow. In 1847, the Great Southern & Western Railway Company opened a line from Dublin to Portarlington and when the Central Ireland Railway Company began to run from Portlaoise to Mountmellick in 1885, almost overnight passenger traffic on the canal stopped altogether.

After the railway reached Portarlington in 1847, the canal's traffic was of bulky, low-value cargoes. Guinness was the main product brought from Dublin, and there was heavy traffic in coal, timber and hardware, and smelly cargoes of manure, tar and hides. The traffic from Dublin was malt, flour, wheat, beet and sugar.

In 1911, the Grand Canal Company fitted a Bolinder semi-diesel engine on one of its formerly horse-drawn barges and, by 1939, all its barges had engines. Twenty years later, the company began to build up a fleet of lorries to carry goods from the canal to customers. The railways had ended passenger traffic on the canals. Nearly a century later, lorries would finish off goods traffic and ultimately cause the closure of the Mountmellick Canal.

Traffic to Mountmelllick ceased in 1940 when the last load of malt from Codd's Maltings was shipped to Dublin, although grain barges from Dublin to Odlums Mill kept the section to Portarlington open.

After the Second World War, traffic on the canal continued to decline. In 1950, the Grand Canal Company became part of Córas Iompair Éireann (CIE), the new transport authority set up by the government, which also took over the lorry services related to the canal. As goods traffic shifted to the roads, the canal trade dwindled to nothing.

The canal officially closed in 1960. Afterwards, in places where it ran through privately owned land, CIE offered to sell sections of the canal to the landowners. Responsibility for those sections of canal that remained unsold was eventually transferred to Waterways Ireland, the cross-border agency with responsibility for all Irish inland waterways.

Between 1969 and 1971, Laois County Council filled in most of the canal between the second and third locks at Tinnakill to make the Portarlington ring road. This stretch of just over 2 miles (3.2 kilometres) is the longest continuously infilled length of the canal.

North-east of Blackhall Bridge, the old canal disappears altogether. The only sign that it ever existed is the railway bridge. Further on, the canal is open, well-preserved and often contains water. There is no open water in the last section, which stretches from the third lock to the terminus at Mountmellick. More than half of this length has been infilled. However, between Tinnakill Lock and Dangan's Bridge, there is a good stretch of open, but dry, canal bed. The infilled canal runs through open fields on either side of Tinnakill Lock. In places, almost nothing remains to show the original line of the canal.

Ordnance Survey of Ireland Grid Reference: 645070 706970

Newry Canal

During Oliver Cromwell's murderous campaign in Ireland in 1649–52, a survey was made for a canal from Portadown to Newry, and Colonel Monk ordered that a 'navigable trench' should be cut. However, despite the order, no work was carried out. Francis Nevil, a tax collector for the government, made a survey and proposal for a similar canal in 1703, but nothing came of that either.

Work on the Newry Canal did not begin until 1731. Thomas Burgh, the Surveyor General, had interests in the collieries at Ballycastle, which would be affected by competition from Tyrone coal becoming more generally available. Burgh died in 1730 and was succeeded by Edward Lovett Pearce, and work began on the canal in 1731. Although Pearce was officially responsible, he was at the time building the new Parliament House in Dublin, so he gave the task to Richard Cassels, a Huguenot, one of his architectural assistants who had first-hand knowledge of canals on the continent. However, he failed to impress the commissioners and was dismissed after three years.

The English engineer Thomas Steers, who had been assisting Cassels with the scheme, took over. Steers was England's first major civil engineer and built many canals, the world's first commercial wet dock, and the Old Dock at Liverpool. He was paid a fee of fifty guineas for carrying out the survey in 1736 and the following year was given a regular contract. Although the canal took up more of his time than he had anticipated, he completed it within five years. On 28 March 1742, the first two vessels, *Boulter* and *Cope*, passed down to Newry from Lough Neagh with coal for Dublin.

The Newry Canal, the oldest summit canal in the whole of the British Isles, had two sections; the inland canal from Portadown to Newry, and the ship canal from Newry to the Victoria Locks, halfway along Carlingford Lough, that now forms part of the border between Ulster and the Republic of Ireland.

The two lengths of the canal were major feats of engineering, as impressive in their day as the Manchester Ship Canal was 150 years later. The inland canal was almost 18 miles (28.9 kilometres) long and had fifteen locks of 44 feet by 15 feet 6 inches (13.4 by 1.6 metres) and between 12 and 13 feet 6 inches (3.6 and 4.1 metres) deep. Ten locks carried the canal up to the 3 mile (4.8 kilometres) long summit level before descending through five locks to Whitecote Point, where it joined the Upper Bann. The locks could accommodate boats of up to 120 tons (121.9 tonnes) burden. Originally, the locks were brick-built but the brickwork began to crumble and was replaced with stone facing from the Benburb quarries in the Blackwater valley. The bottom of the lock chambers was fendered with 2-inch-thick (5-centimetres) deal wood.

The Point of Whitecote, a short distance south of Portadown, is the start of the inland canal, where it runs between the River Cusher and the Upper River Bann. The length to Lough Neagh is to the north and is supplied by the River Bann, entering Lough Neagh, sometimes called 'Ulster's Inland Sea', at Bannfoot. The length to the Bannfoot is roughly 10 miles (16 kilometres). On the Bann and Lough Neagh, the lighters which used the canal were rigged with masts and sails.

The canal was built, like many others, principally because of coal. Coal seams had been discovered at Coalisland in County Tyrone. In order to transport the coal rapidly and efficiently to the capital, Dublin, a canal was necessary. It is assumed, although without firm evidence, that the town was named because of its coal deposits. Although a single barge drawn by one horse could carry as much as 70 tons (71.1 tonnes), it was slower than a train of waggons full of coal, but much more efficient overall. Another attraction of the plan to transport coal from the coalfield near Coalisland, by way of Lough Neagh and the port of Newry, was that the same boats could hoist sail and continue to Dublin by sea.

At the bridge at Moneypenny's Lock is the lock-keeper's house, a two-storey building with a gable facing the canal. Beside it is a 'bothy' which acted as a warehouse, stabling for eight canal horses, and lodgings for lightermen. Members of the Moneypenny family were lock-keepers there for eighty-five years.

Many of the lighters were built in Portadown by two main boat builders, the Portadown Foundry and Bright Brothers. The lightermen were clearly romantics, naming their boats after their wives or girlfriends. Frank Neill's was called *Nora*, Billy McCann's was the *Flora*, James Neill's was *Emma* and T. M. Grogan's was *Edith*. Cargoes included linen, clay products, farm produce and imported goods for businesses in Portadown, as well as coal. Besides being reasonably successful in terms of cargoes carried, the Newry Canal offered a passenger service which operated from Portadown to Newry. This was another business venture started by a Quaker entrepreneur from the area.

When the railway first came, the morning train from Belfast connected with the passenger-carrying 'packet' boats three days a week. The journey by packet took about four hours, which meant that business people could go in the morning, transact their business, and return in a single day. The business success of Portadown and Newry was the direct result of the canal, and Newry became the fourth most successful port in Ireland.

Although originally built to supply coal from the Tyrone coalfields, the coal from Coalisland proved to be of poor quality and was expensive to mine. The real threat to the viability of the canal, though, was the railway which ran (and still runs) alongside the canal virtually the whole way to Newry. The last vessel to use the inland canal was a pleasure yacht in 1937, and it was finally abandoned in 1949. The length through Newry itself continued to be used until 1956 for goods to be stored in the warehouses, and the ship canal continued to trade until 1974 when the port of Warrenpoint was established. Without Newry as an outlet to the sea, the value of the inland canal was limited, so, in 1759, improvements began under John Golborne, who made a small cut parallel with the river, which was hopelessly inadequate. Thomas Omer replaced it with a true ship canal 1.75 miles (2.8 kilometres) long and of more generous dimensions, with a sea lock at Lower Fathom measuring 130 feet by 22 feet (39.6 metres by 6.7 metres) for coasters. Construction was finished in 1769. The ship canal did a great deal for the development of Newry.

The Commissioners of Inland Navigation were disbanded in 1787, and the canals came under local control. However, in 1800 the Directors-General of Inland Navigation were appointed and between 1801 and 1811 both canals were extensively repaired. The locks on the inland canal were rebuilt and enlarged for vessels 62 feet by 14 feet 6 inches (18.8 metres by 4.4 metres), the same size as those on the Tyrone Navigation.

Even after the repair programme the ship canal section was still unsatisfactory because of silting. Steamships were now becoming more common, so in 1830 the engineer John Rennie recommended an improved natural channel from the sea lock down to Warrenpoint. In fact, the ship canal was extended even further downstream, to Upper Fathom, where the Victoria Lock was built, measuring 220 feet by 50 feet (67 metres by 15.2 metres). The canal, which was now 3 miles (4.8 kilometres) long, was completed in 1850 with a new basin at Newry, while below the lock the natural channel was improved down to Warrenpoint.

The work had been done by a local company, the Newry Navigation Company, which had been created in 1829. After 1858, the inland canal did less well due to the railway, and its condition deteriorated. By this time, too, Belfast had replaced Newry as Ulster's chief port, although Newry tried hard to compete, deepening and widening the natural channel in the 1880s to take 5,000-ton ships. The Newry Navigation returned to public control in 1901, when the company was replaced by the Newry Port & Harbour Trust. Further improvements were made to the harbour and ship canal, and efforts made to keep the inland canal in working order. However, there was no traffic on it after 1938, and it was abandoned in 1949 as far as Newry, the town section to the Albert Basin being abandoned in 1956. In 1958, the Dublin road swing-bridge was fixed so that vessels could no longer enter the inland canal, although the channel was kept as a feeder to the harbour and the ship canal, which remained open until 1976 when the large harbour, planned at Warrenpoint to handle traffic for the new city of Craigavon, was opened.

Ordnance Survey Grid Reference: SB 16853 84372

Newry Canal just after dawn. Image used by permission of John Muldoon.

Ships on the canal at Merchants Quay in Newry, alongside a nice brace of merchants' premises, in the nineteenth century. Courtesy of the National Library of Ireland.

Park Canal

The 1-mile-long (1.6-kilometre) Park Canal in Limerick was built in 1757–58 as a commercial waterway for transporting goods from Limerick on to the Shannon, and from there on to the Grand Canal and eventually to Dublin. It bypassed the loop in the Shannon immediately north of Limerick. Its engineer was William Ockenden.

Using primitive tools and limited engineering techniques, the first part of the Park Canal to be built was the First Lock of the Old Canal Harbour off Clare Street. The canal went through Bartlett's Bog below Troy Lock. For just a few pence a day, supplemented by payments from the Pery Charitable Loan Fund, the workmen used buckets and spades to shift tons of sodden earth. Recent dredging has shown that the canal was puddled with local yellow clay. The digging of the canal cost £401. The next length was an even greater challenge, involving cutting through a seam of solid rock. The head of the canal was reached in 1758 but the whole canal was not formally opened until 1861.

The structures relating to the canal include the First Lock and Park Lock, both constructed of limestone masonry, and two bridges, both single-arched. The First Lock Bridge was built with red brick by a Dutchman called Uzuld (possibly Henry), who was a merchant rather than an engineer, while Park Bridge is built of limestone. Uzuld also designed the paper mills which stood to the west of the canal. There is also an original triangular milestone between the Shannon and Park Lock with the distances to Limerick and Killaloe. The canal ran in a straight line from the Abbey River to a junction with the Shannon.

As an early canal, the gradient of the banks in the upper section is steeper than later engineers would have chosen. The majority of structures relating to the canal and buildings along it are between Lock Quay and Park Bridge. Further east were the old Guinness warehouses (which have now been demolished). The lock gates at Lock Quay were replaced in the early twenty-first century as part of the Limerick Main Drainage Project.

Edmund Sexton Pery, the Member of Parliament for Limerick, Sir Henry Hartstonge and members of the Maunsell family saw, in the opening of the canal link with the Shannon, an opportunity to develop the commerce of Limerick, not least with a view to provide the poor of the city with some income. However, the amount which these gentlemen were able to raise was not as much as was clearly needed.

In 1767, under an Act of Parliament, the Limerick Navigation Company was established, with a grant of £6,000 to be administered by Limerick Corporation. In 1813, the British government bought out the company and vested its powers in the Directors-General of Inland Navigation. The low level of industrial activity in the city meant that traffic did not grow, although waterborne traffic was cheap at 3d per mile and 2d per lock. Those cargoes that were carried included domestic goods such as bread and milk and whiskey, as well as coal, timber, stone, brick, lime and manure. Turf was particularly important during the First World War, when ships carrying coal from England ran the gauntlet of U-boats operating in the Irish Sea.

In 1765, the lock mills by the canal were powered by waterwheels with water from the canal. The mill owners, the principal one of whom was John Norris Russell, paid the Shannon Navigation Company £40 a year for this.

Hundreds of families in Limerick made their living from various commercial activities related to the canal. The canal manager's office was near the original site of the First Lock, of which only the red walls remain. The First Lock House was also the head office of the Canal Management Committee.

In the eighteenth century, the barges used on the canal were flat-bottomed and were from 6 to 8 tons (6.09 to 8.1 tonnes) burden, with masts that could be lowered and square sails. However, by 1913 barges like the *Athy* and the *St Patrick*, capable of carrying 40 to 45 tons (40.6 to 45.7 tonnes), traded into the city carrying coal and, inevitably, Guinness. They carried sand and marble from Ballysimon and Garryowen out to seagoing ships in the offing.

With the growing use of motor lorries after the First World War, and the building of the hydroelectric generating station at Ardnacrusha, the Park Canal became obsolete. Most commercial carrying on it finished in 1929, although traffic did resume for a couple of years after that, partly because of the rebuilding of Park Bridge and partly because the Grand Canal Company felt that the new route through Ardnacrusha and down the Abbey River was too dangerous. The last Guinness barge to use the canal was in 1960.

Originally, the port of Limerick was near the confluence of the Abbey River and the Shannon at King's Island, but today the port is located further downstream on the Shannon. Historically, Limerick made products from agricultural produce from the surrounding area, because of its position as the first major port along the River Shannon.

There has been much debate about restoring and reopening the canal this century, but hard times economically, together with practical difficulties, have led many people to question the idea. For example, although there is a reasonable depth of water in the upper section of the canal, overall the level is well below what it was before the building of the power station at Ardnacrusha. The first 10 cubic metres (13 cubic yards) per second of water are sent down the original course of the river, by way of O'Briensbridge, Castleconnell and Plassey, which is why the level in the upper section of the canal is fairly high. The next 400 cubic metres (523.1 cubic yards) per second of water go through the power station and, if there is any water left after that, it too is sent down the canal's original course. The lower overall level means that the quays are well above the waterline, except in floods, and Limerick City Council has made the quays higher, making the canal less usable.

Relatively few pleasure boats go to Limerick from the Shannon. For them to go into the canal, they need to be persuaded that it would be worthwhile. As it is now, the lower section of the canal offers views of nothing more picturesque than high quay walls surrounded by derelict buildings.

The path along the canal has recently undergone extensive redevelopment to form a cycling and walking route between the city and the University of Limerick.

Ordnance Survey of Ireland Grid Reference: 558946 657692

Plassey–Errina and Killaloe Canals

Before the Ardnacrusha hydroelectric power station was opened in 1929, there was another waterway route, as well as the Shannon, from Killaloe to Limerick, which had three canal sections and two river sections.

The navigation was built by the Limerick Navigation Company. Construction was difficult, lengthy and costly, and after forty years the Directors-General of Inland Navigation took it over and completed the work. The Killaloe Canal is situated on the River Shannon and the two are separated by a narrow wall constructed in the late eighteenth century to allow vessels to bypass an area of rapids on the Shannon, providing for the uninterrupted passage of vessels between Limerick and ports south of Killaloe. Most of the Killaloe Canal was submerged under the new lake, the 'flooded section' south of Killaloe created as part of the works of the hydroelectric station.

The Killaloe Canal originally had three locks, two of which were drowned when Ardnacrusha was built. Below that, boats used the river for the next section, which brought them past the village of O'Briensbridge. Then, about 1.2 miles (2 kilometres) further downstream, they turned right into the Plassey–Errina Canal, although it was also possible to go a little further downstream to a quay at the upper end of the village of Castleconnell.

The Plassey–Errina Canal had six locks, two of them doubles. The uppermost lock, at Errina, was originally a triple-chambered lock, unique in Ireland, but it was later converted to a double. The canal rejoined the Shannon at Plassey, where the University of Limerick now is. At first the boat horses were ferried across the river, although a bridge was later built.

The building of the hydroelectric station, and the weir at Plassey Villa, had not only flooded an area below Killaloe but had also blocked access to the stretch of river that included O'Briensbridge, Castleconnell and the entrance to the Plassey–Errina Canal.

Marked on the Clare Grand Jury Map of 1787 and the 1842 first edition Ordnance Survey map, the towpath now forms part of the Lough Derg Walkway. The Plassey–Erina Canal runs from the Shannon to the north-east of Clonlara for a distance of 5.5 miles (9 kilometres) before re-entering the Shannon at Plassey, north-east of Limerick city. The canal formed part of the Shannon Navigation and was built, in around 1770, to bypass the falls at Castleconnell and Doonass.

Newtown Lock is the only lock on the Plassey–Errina Canal that has a date – 1792 – carved on it. The lock is of fine ashlar limestone set in lime mortar, but the gates have been removed. Next to the lock is a single-storey lock-keeper's house. The single-arched Errina Bridge carries a local road over the canal, which at this point is in a deep rock cutting.

The Killaloe Canal also forms part of the Shannon Navigation Scheme and, similarly, was built in the 1790s to bypass the falls and eel weirs at Killaloe. The southern section of the canal was flooded when the Shannon hydroelectric scheme was constructed, including Cussane Lock, although the northern section is still navigable.

The pier head at Killaloe at the northern end of the Killaloe Canal is constructed of square limestone blocks. There is a dry dock there which is used for boat repairs; lock gates give access to the dry dock. Johnson's House, near the pier head, was probably originally associated with the canal.

The Killaloe Canal itself is stone-lined throughout, and has three locks – Killaloe Lock, Moys Lock and Cussane Lock. Structures such as goods sheds, slipways, cranes, mooring posts, locks and cast-iron mooring rings are all still intact. Killaloe Lock was built immediately north of Killaloe Bridge. The lock-keeper's house, which formerly stood between the canal and the bridge, was demolished in the early 1990s and the Shannon Heritage Centre has been built on its site. There is a stone single-arched road bridge over the canal.

At Shantraud, the former slipway and canal quay, on the west side of the canal, is no longer in use. The slipway was used to hoist boats up to a dry dock for repair.

From Errina to O'Briensbridge, the towpath has been cleared and is now the 'Old Barge Way'. Looking across the Shannon it is possible to see to the entrance of the old canal system into Macnab's Bog, Montpelier, which was used to supply the Limerick Distillery with peat.

Ordnance Survey of Ireland Grid Reference: 569767 673277

Sun trying to break through the early morning fog on the Killaloe Canal. Image used by permission of Cormac Costello.

Rockville Navigation

The Rockville Navigation was a short canal built on the Rockville Estate, County Roscommon, the home of a branch of the Lloyd family. The estate had originally belonged to the Blackburns, but came into the possession of the Lloyds following the marriage of Owen Lloyd of Lissadorn to Susanna Blackburn in 1740. Colonel Owen Lloyd held a considerable amount of land in the 1850s. However, by the 1860s most of his estate was in the hands of the Court of Chancery (which had jurisdiction over all matters of equity, including trusts and land law), although it is not known why. Rockville House had been valued at £45 in the 1850s.

On 1 December 1906 the *Irish Times* reported that Major and Mrs Lloyd of Rockville Hall had given a ball to their tenants, celebrating the coming of age of their son and heir. Major Lloyd, who at that time was a Justice of the Peace and a Deputy Lieutenant of the county, died in 1912.

In 1917, the young Mr Lloyd was appointed a magistrate for County Roscommon. The following year, he sold the Rockville Estate to George Frayne of Ballaghderreen, after his family had owned it for more than 200 years. The house was demolished in the second half of the twentieth century.

The navigation still exists, and, although it has not seen a boat for a considerable length of time, the water level appears to be quite high, with a definite flow. The date carved on the bridge – June 1765 – places the canal in the first period of the building of Irish waterways in which, for example, the first man-made cuts on the Shannon were made in 1755–69. The high stone bridge has an arch which is about 15 feet (4.5 metres) wide, while the top of the parapet is about 16 feet (4.8 metres). Its clearance of almost 11 feet (3.3 metres) over the present water level makes it comparable to bridges on the Grand Canal, although here there is no towpath under the arch.

The navigation joins Lough Nahincha, a bog lake with an island, which is about 16 feet (4.8 metres) deep. On one side of the lough there were two small canals, one of which was the navigation, while the other seemed to run to an orchard. The 1914 Ordnance Survey (OS) map (surveyed the previous year) showed a boat house at the end of the latter.

The navigation seems to have been largely forgotten since three early writers on Irish inland waterways – Harry Rice, John Weaving and Hugh Malet – mentioned it. Only Edmund Sexton Pery actually described boating on it, and he provides evidence that at least some parts of it were artificial cuts.

Other OS maps in the nineteenth and early twentieth centuries support Malet. The straight cut with the bridge and two short cuts near Rockville House, linked to the fish pond and an orchard, are marked as canals. It is difficult, however, to distinguish between canals for navigation and drainage channels. On the nineteenth-century OS map, the channel linking the straight cut to the Lough was a very winding one, but was shown as a clean curve on the 1914 map, although there is no means of knowing whether this was done to improve it for navigation or drainage.

The evidence on the ground would seem to support Hugh Malet's judgement. Although it seems most likely that the navigation was cut by a member of the Lloyd family, there is no documentary evidence of who commissioned it, what it cost, who actually built it (although it was most likely to have been men who worked on the estate), or why it was built. Brian Goggin speculates that boats used on it may have been wooden 'cots', similar to those used on the Shannon and other waterways, which were quite wide and up to 35 feet (10.6 metres) long.

If it really was a navigation, cargoes may have included turf, potatoes and other agricultural produce, and manure. For what it is worth, a local priest, who remembered as far back as the mid-1940s, said that there had been no traffic on the waterway.

Ordnance Survey of Ireland Grid Reference: 595538 791212

Rockingham Canal

Immediately to the south of Lough Key, which is connected to the upper reaches of the Shannon by the meandering River Boyle with one lock, is the Lough Key Forest Park, which opened in 1972. The park covers 865 acres (350 hectares) and was formerly part of the Rockingham Estate. The Moylurg Tower, a five-storey concrete viewing tower, about which opinion is divided – some seeing it as iconic modern architecture and others as an eyesore – was built on the site of the house in 1973.

The King family had acquired the land around Lough Key in the seventeenth-century Cromwellian Settlement, and changed the name of the area from Moylurg to Rockingham. In his book *Green and Silver*, L. T. C. 'Tom' Rolt describes coming across the mouth of a canal where it joined the lough, and a little distance away discovering the ruins of a lock. He wrote,

> These canals, we found out later, were not merely ornamental. They once combined beauty with utility by enabling turf to be brought by boat from the neighbouring bogs to a quay on the lakeside immediately below the house. Thence the turf was conveyed to the domestic quarters by means of an inclined subterranean passage beneath the terraces. I have seen private estate railways such as that belonging to the Duke of Westminster at Eaton Hall, but this was the first time I had seen a private canal system constructed for anything other than a purely ornamental purpose.

There are several short canals, but the one with the lock is east of where the house once stood. Out of all the 'turf canals' that are known in Ireland, this is the only one with a lock. There was also a small harbour on Lough Key, and close to it a group of farm buildings, suggesting that the canal may have carried agricultural produce, as well as turf. The canal, which is crossed by a balustraded ornamental stone bridge, is today shallow and overgrown.

Ordnance Survey of Ireland Grid Reference: 582816 803424

Roscrea Canals

John Birch's whiskey distillery, at Birchgrove, Roscrea, began operations in about 1780. In a very short time it had built up a considerable reputation. In 1815, *Atkinson's Tourist Guide* said: 'The principal commercial feature of Roscrea, a town of considerable extent and of some trade is that of whiskey manufacturing which supplies the dealers in Irish Spirit with considerable quantity of that favourite liquor.'

The distillery was quite large and some of the outbuildings can still be seen at Birchgrove. When it was working, a canal 4 miles (6.4 kilometres) long was dug from Monaincha to Birchgrove, along which barges full of turf – the fuel used for the stills – were drawn by donkeys. The turf came from a bog to the east and south of Racket Hall, which began life as a gentleman's residence, but is now a hotel. The Ordnance Survey (OS) maps of 1838 to 1840 show a small canal network in the bog near Roscrea.

Lewis's Directory said:

> Corbally Co. Tipperary. A private canal, about four miles in length, has been constructed, from which are several branches, one for conveying turf to the distillery at Birch Grove, and another to the Rathdowney road leading to Roscrea, and partly supplying the latter town; all run into the bog of Corbally, in which is a lake about one Irish mile in circumference. A considerable portion of the bog has been reclaimed by Messrs Birch, and is now in a high state of cultivation.

Although some of the bog appears to have been drained since the OS map was drawn, it is still quite large, with many small watercourses to the east and to the south of Birchgrove, although they may have been drains rather than canals, or possibly combined the two functions. It is not possible to identify the private canal mentioned by Lewis. The vessels used may have been cots of some kind, the generic boat type of the district, which are flat-bottomed but vary greatly in size and function.

For reasons that are not known, Birch's Distillery got into difficulties between 1810 and 1820. John Birch was trying desperately to make the business profitable once more. In an attempt to make the process of distilling more efficient, he developed a wooden jacket for stills that was used as a steam case. His success was short-lived, however, because it attracted the attention of some of the powerful distillers in Dublin. They informed the Board of Excise, suggesting that illegal poteen brewers would use the idea to hide the steam from their stills, and the invention was rapidly banned. Birch had to stand by and watch as his business slowly died. To add insult to injury, in 1850 he received a large fine for the evasion of taxes, which led directly to the closure of the distillery.

In 1837, Lewis recorded that the house called Birchgrove, which still exists, belonged to John Birch.

Ordnance Survey of Ireland Grid Reference: 613622 689283

Ulster Canal

In 1778, a canal was proposed to run from Ballyshannon to lower Lough Erne. Its estimated cost was £32,000, but it was seen merely as part of a more ambitious plan to create a navigable waterway connecting the ports of Belfast and Coleraine with the Shannon, allowing access to Limerick or Waterford, but this never came to fruition. An additional link from the Ulster Canal to Enniskillen, Belturbet and Ballyconnell was estimated at a further £8,000. A possible future link from Ballyconnell to Ballymore, along the Woodford River valley and on to Lough Scurr and the River Shannon at Leitrim, was also suggested but not costed.

It was envisaged the Ulster Canal would be an important section of a great waterway crossing Ireland from east to west, competing with the similar east–west link formed by the Grand Canal and Royal Canal through Dublin. The parliament in Dublin provided funding in 1783, and a section of the canal was constructed between Ballyshannon and Belleek, with Richard Evans, the engineer for the Royal Canal, overseeing the work, which included a lock at Belleek. However, funds ran out in 1794 and construction came to a halt.

In 1801, the Directors-General of Inland Navigation asked Evans to prepare an estimate of what it would cost to complete the canal, although no action was taken. But by 1814, faced with unemployment in the area, the Directors-General decided that a canal from Lough Neagh to Lough Erne – part, at least, of the original plan – would provide jobs for local people. John Killaly, probably the most significant Irish canal engineer, who had worked for the Grand Canal Company and then as an engineer for the Directors-General, was asked to survey a route. He produced his report in February 1815, estimating that a canal rising through six locks from Wattle Bridge to a summit near Monaghan and then descending another sixteen locks to reach Lough Neagh would cost £233,000.

It would be 35.5 miles (57.1 kilometres) long, including a branch to Armagh. Killaly, for some unfathomable reason, proposed to make the locks of a similar size to those on the Royal Canal, 14 feet by 76 feet (4.3 metres by 23 metres), which would accommodate boats up to about 13 feet 4 inches (4.1 metres) wide, despite the fact that the vessels that already used Lough Neagh, the Lagan Canal, the Newry Canal and the Coalisland Canal were 14 feet 8 inches (4.5m) wide, and would therefore be unable to use his canal.

A public meeting was held at Monaghan in February 1817, at which a group of landowners and businessmen offered to put up two thirds of the cost of the canal. Despite that, the Directors-General took no action, and the scheme was not started. However, the proprietors of the Lagan Canal saw the proposed canal as a way of increasing traffic and revenue on their canal. There was growing public support for the idea, and eventually a public petition was presented to Parliament for a scheme, which was very similar to Killaly's of ten years earlier. The government doubted it would see a return on any money they might advance, meaning that the Directors-General were powerless to fund the project. Eventually, in 1825, a private company was authorised to build the canal, which was by then estimated to cost £160,050 – £70,000 less than Killaly's estimate – because a new survey proposed a line that would need only eighteen locks.

The company then applied to borrow £100,000 from the Exchequer Bill Loan Commission, a body that had been set up under the Poor Employment Act of 1817. The leading engineer Thomas Telford was asked to go to Ireland to look over the plan and estimate, which he duly approved. However, there was a dispute about what interest rate should be charged on the loan, and it took three further Acts of Parliament before a loan of £120,000 was agreed.

There was then a host of further problems. The contractors, Henry, Mullins & MacMahon of Dublin, had been awarded the contract to build the canal in 1832. Telford, who was nearing the end of his life, now decided that the design had serious flaws and that a new survey was required. This increased the number of locks needed to twenty-six, and the contractors were asked for a new estimate. However, no agreement was reached, and they withdrew from the project, being replaced by William Dargan. To complete the succession of catastrophes, John Killaly, the local engineer, died that year. It had been decided to reduce the width of the locks, but it is not known whether the decision was his or Telford's. Whatever the truth of the matter, the locks were built 12 feet (3.7 metres) wide, making through traffic impossible except in specially built boats.

William Cubitt succeeded Telford, after he also died, in 1834. The canal, which, it should be remembered, had first been proposed in 1778, was finally completed in 1841. Nineteen locks descended to Lough Neagh from the summit and, in the other direction, seven descended to Lough Erne. Water was supplied from the Quig Lough near Monaghan, which had been enlarged to act as a reservoir. Inexplicably, the final lock at Wattle Bridge was only 11 feet 7 inches (3.6 metres) wide, which made it the narrowest lock in the whole of Ireland. The project had cost over £230,000 although, remarkably, that was some £3,000 less than Killaly had estimated in 1815.

All in all, the canal had been ill-considered in the first place, and was an abject failure commercially, contributing to the collapse of the Lagan Navigation Company, which had been refused permission to abandon it when it could no longer afford the cost of maintenance. The canal failed to generate a significant amount of trade, partly because the water supply was inadequate, and because goods had to be transshipped at either end into and out of the narrower boats that were needed to navigate it.

The link to the River Shannon, to generate through traffic, that had been envisaged had not been built, and was never likely to be until the canal earned worthwhile profits. The Lagan Company was unable to repay any of the loan made by the Exchequer Bill Loan Commissioners so, in 1851, the Board of Public Works took control of the canal. After some largely cosmetic repairs, it was leased to William Dargan, who had built most of it, and who ran the only significant carrying on the canal. In 1858, the Ulster Railway reached Monaghan, and three years later the canal, having been allowed to fall into a ruinous state, was impassable.

In an attempt to recoup its considerable losses, the Dublin government took control of the canal for a second time in 1865, officially closed it, and spent £22,000 on repairs over a period of eight years; a clear case of throwing good money after bad. The government's priority had been to provide an adequate supply of water but when the canal reopened in 1873, it was obvious that this had not been achieved. The continuing cost of maintenance greatly exceeded the revenue generated and, as the summit was mostly unnavigable, the little traffic that existed was limited to the Lough Erne end, on top of which there was only a passable depth of water for eight months of the year.

However, there was some increase in traffic in 1880, when the secretary of the Lagan Navigation Company, W. R. Rea, set up a new carrying company using smaller boats – for this common sense decision alone, he deserved to succeed. The government vaguely suggested that it would provide assistance to any company interested in taking over the canal. A series of negotiations with the Lagan Canal Company took place, but the government failed three times to pass a Bill to authorise the company to buy the canal. Eventually it suggested that the Lagan Canal should obtain a private Bill to take it over, which it succeeded in doing in 1888.

The House of Lords, however, had removed a clause from the Bill that would have allowed the Lagan Canal Company to close the Ulster Canal after ten years, and as a result the company was saddled with a liability with no apparent end. The company had to spend huge amounts on maintenance while earning a trivial amount of income. Profits from the Lagan Canal and the Coalisland Canal, which it also owned, were swallowed up by the legal requirement to keep the Ulster Canal open, and the company never really recovered from its toxic acquisition.

The last boat entered the canal in 1929, and on 9 January 1931, the company finally obtained permission to abandon the section of the canal in Northern Ireland. Relief from the liability for maintenance was granted on 15 April 1931. However, that was not the end of the story because an Order of Abandonment was refused by the government of the newly established Republic of Ireland.

In the 1940s, Monaghan County Council was granted a Judgement Mortgage on the canal in lieu of unpaid rates. The Lagan Company seized the opportunity and offered to give the canal to the council in compensation for the money it owed. The council declined, however, having been advised that it would not have the financial resources to pay for the upkeep of bridges and other structures. The Lagan Navigation Company was dissolved under the Inland Navigation Act (NI) 1954. This then led to a period in which various government and local government bodies acquired stretches of the canal under different pieces of legislation.

The canal follows a more-or-less straight course from south-west to north-east from Wattle Bridge, on the River Finn, to Charlemont, where it joins the Blackwater. There were two locks close to the River Finn, two beyond Clones, and three near Smithborough, where the summit level was reached. The summit pound, less than 6 miles (10 kilometres) long, descended two locks before reaching Monaghan, and a flight of seven shortly after the town. The border between the Republic and Northern Ireland crosses the canal below that flight. There are two isolated locks near Middletown, and then a level section before the canal reaches a gorge to the west of Benburb. Constructing the canal through rock presented severe problems for the builders, and a further six locks had to be built in difficult terrain. There is one more lock above

Blackwatertown and then the final lock below Charlemont before the canal makes a junction with the River Blackwater.

There has been a long-standing campaign to have the Ulster Canal reopened to navigation, and in 2013 restoration work began on the first 7 miles (11.2 kilometres) of the canal in the Republic. The intention is to preserve the existing route of the canal and its towpaths, together with road bridges. Two existing bridges over the canal and a double lock will be restored at Gortnacarrow, allowing vessels from the River Finn to reach the canal section, and a marina will be built. But it is unlikely that the remaining 39 miles (62.7 kilometres) of the canal will be reopened in the foreseeable future.

If the Northern Ireland Executive and the government of the Republic both agreed with the plan, it would see vessels cruising from Coleraine on the north coast, through Lough Neagh and Lough Erne, into the Shannon and then through the canal system in the Republic to Dublin, Limerick and Waterford: the 'missing link' of the waterways in Ireland.

Ordnance Survey Grid Reference: SA 76379 95463

SCOTLAND
Aberdeenshire Canal

The Aberdeenshire Canal, designed by John Rennie, ran from the port of Aberdeen to Port Elphinstone, Inverurie, and opened in 1805. Port Elphinstone, now a suburb of Inverurie, is on the right bank of the River Don. It took its name, and during the first half of the nineteenth century derived its importance, from being situated at the north-west end of the Aberdeenshire Canal. The canal was originally intended to be part of a scheme linking Aberdeen to Monymusk, going via Inverurie, where there would be a branch along the course of the Urie Glen to Insch. Captain George Taylor conducted a survey which confirmed that the scheme was feasible. However, only the first section to Inverurie was actually built. As constructed, it was only 17 feet wide with a depth in the channel of 3 feet, although Rennie had proposed a 27 foot width and a depth of 4 feet. However, during the first six years of operation the dimensions were gradually increased over much of the canal to a maximum width of 23 feet and depth of 3 feet 10 inches.

An Act of Parliament, obtained on 26 April 1796, created 'The Company of Proprietors of the Aberdeenshire Canal Navigation', authorising them to raise £20,000 in £50 shares. Work began in 1796, but five years later the proprietors ran into financial difficulties due to only £17,800 having been subscribed. All of the capital had been spent, leaving the company in debt. A Bill was put before Parliament, and a new Act was passed on 24 March 1801 in the hope that it would attract more investors.

The canal eventually opened in June 1805, and *The Aberdeen Journal* reported:

> The *Countess of Kintore*, handsomely decorated and fitted up by Captains Bruce and Freeman … proceeded to Kintore, where they were met by the Magistrates, and other inhabitants of that burgh. On their approach towards Aberdeen, they were joined by several parties of Ladies, who were highly pleased with the novelty of the navigation through the locks; while several thousands of the inhabitants, crowding on the banks and bridges.

However, if the proprietors heaved a sigh of relief, it was premature. Fourteen of the locks failed within weeks, and the canal was shut until 1806 while the masonry was renewed. A third Act of Parliament was passed on 13 March 1809, authorising the company to raise a further £45,000 to complete the canal.

By the end of 1806, the rebuilding of the locks was finished and traffic on the canal resumed. Boats carried a variety of cargoes from the agricultural hinterland of Aberdeen, stone from a number of granite quarries, and the products of paper mills. There were milestones at half-mile intervals, in order to calculate the tolls for the carriage of goods and passengers. A passenger boat was introduced in 1807 and ran three times a day for most of the year from 1 April, but only once a day between October and December. Passengers were only carried on the top section of the canal, and disembarked at a building called the Boathouse, situated above the five St Machar Drive locks. The prevailing north-easterly winds, and ice, dictated that the canal was only open from 1 April to 1 December.

Another handicap for traders on the canal was that, initially, it did not connect directly with the harbour in Aberdeen. All goods had to be transshipped from seagoing vessels to barges until a tidal lock was constructed in 1836. By that time, however, this did little to revive the profitability of the canal, which had never realised its potential.

The company was permitted to take water from any watercourses within 2,000 yards (1,828.8 metres) of the canal and from the River Don, along the valley of which the canal ran from Aberdeen for a total of 18 miles (29 kilometres). It terminated just south of Inverurie. The terminal basin was named Port Elphinstone, after Sir James Elphinstone, who was a major shareholder and lived at Logie House near Oyne and Pitcaple.

On the day the canal opened in 1805, the occasion was marked with an inaugural voyage by the barge *Countess of Kintore*. Landowners, shareholders, magistrates and other important people boarded the boat in Inverurie for Aberdeen and were entertained by the band of the Stirlingshire Militia. The journey took seven and a half hours.

There were seventeen locks, each 57 feet by 9 feet (17.3 metres by 2.7 metres), all in the first 5.5 miles (8.8 kilometres) between Aberdeen harbour and Stoneywood, with fifty-six road bridges, twenty culverts and five aqueducts. A towpath ran along the right-hand bank heading north-west, all the way from Aberdeen, and any wharfs and moorings were on the left, so that the passage of other vessels would not be impeded. From Stoneywood, the canal ran level to Port Elphinstone. The fall from Stoneywood to the low-water mark at Aberdeen was 168 feet (51 metres). Throughout, the canal was 23 feet (7 metres) wide, with an average depth of 3 feet 9 inches (1.1 metres).

For most of its route, the canal was cut through granite, some of which was quarried and transported to London for building. Revenue from tolls rose from £311 in 1807, the first full year of operation, to £3,062 in 1853, which was the last full year of operation.

In 1845, the shareholders met to discuss the construction of a railway from Port Elphinstone to Huntly. The meeting resolved to buy the canal and to use its route for a railway to be called the Great North of Scotland Railway. The canal stayed open for use until the line to Huntly was finished, then was gradually abandoned in the direction of Aberdeen. The idea was to use the canal bed for the railway, although, where the canal had followed the contours of the land too closely, the railway was able to take a straighter line. These divergences allow what remains of the canal to still be followed today. At the Inverurie end there is a stretch that is still in water, having been retained for use as a water source for a paper mill which has long since closed. Some other remains of the canal can be seen in Aberdeen. At Woodside, another canal bridge is used by a road, with nothing now running beneath it. In Old Aberdeen there is still a Canal Street.

A bridge spanning the canal lies just north of Great Western Road, at the entrance to Station Road, Woodside. This bridge is a rare survivor, as it spans the only section of the canal in water.

Milestone 14½ still survives embedded in the garden wall of 'Arden Lea', a house previously known as 'Canal Cottage', while the adjacent petrol station stands on the site of one of the canal wharfs.

The basin at Port Elphinstone has completely disappeared, but a plan of part of the Great North of Scotland Railway shows that it was narrow and oblong in shape, with a rectangular north-eastward extension at its north-western end.

Ordnance Survey Grid Reference: NJ 77292 20446

Black Wood of Rannoch Canals

The Black Wood of Rannoch, 'Coille Dubh', is one of the largest surviving areas of the original Caledonian pine forest, which thousands of years ago stretched across Britain and Europe. The wood has been described as a living, growing monument, with some trees thought to be about 400 years old. The Black Wood lies to the west of Dall and is about 3 miles (4.8 kilometres) long. Six miles (9.6 kilometres) west of Kinloch Rannoch, within the South Rannoch Forest, it lies on the southern shore of Loch Rannoch.

The remains of two 200-year-old canals in the wood have been identified, all that survives of an ambitious plan to exploit the forest. Many people think that the Black Wood has remained untouched since time began, but the reality is that it has been shaped by man over many centuries. Over hundreds of years it has provided building materials, fuel and charcoal for use in iron smelting.

In the late seventeenth and early eighteenth centuries, a plan to send the timber to more distant markets resulted in the canals being dug. The venture eventually did not succeed, but the remains of the canals are still visible in the middle of the Black Wood. The logs from trees felled in the wood were sent down chutes into the canals to be floated down to Loch Rannoch, rather than being carried on boats, to a sawmill at Carie that was operated by the Forfeited Estates Commissioners.

In the Second World War, the Canadian Forestry Corps, which consisted of lumberjacks recruited for their skill and was not a military unit, also felled timber at the eastern end of the Black Wood.

The Wood was purchased by the Forestry Commission in 1947 and has been allowed to regenerate.

Ordnance Survey Grid Reference: NN 60166 55100

Borrowstounness (Bo'ness) Canal

From the late 1500s, if not earlier, ships landed cargoes at the 'Ness', where the harbour of Bo'ness later stood. During the 1600s, the population grew steadily as the village of Kinneil declined in size and importance and Bo'ness replaced it as the major centre. In the mid-eighteenth century the town was said to be Scotland's second most important port. When the Forth & Clyde Canal was being built, its proprietors chose Sealock (now Grangemouth) in preference to Bo'ness as the eastern end of their canal. Attempts were made by the merchants and ship owners of Bo'ness to have the Forth & Clyde Canal terminated in their town. They argued that Sealock was an unsuitable place for entry into the Firth, as seagoing vessels had to lie offshore, sometimes for days, waiting for the tide and a suitable wind before they could get in. Bo'ness, on the other hand, they claimed, possessed great natural advantages. However, Sealock was chosen, much to the disappointment of the merchants of Bo'ness. What weighed decisively in favour of Grangemouth is not entirely clear, but it is said locally that it was the influence of interested parties in Grangemouth. Of course, if Bo'ness had been chosen, the inhabitants of Grangemouth would, no doubt, have made similar accusations.

The merchants and ship owners of Bo'ness clearly realised that the Forth & Clyde Canal, when it was finished, would severely curtail, if not completely put an end to, their trade with Glasgow. Carried on by means of packhorses and carriers' carts, this had been very extensive; it was not unusual to see fifty carts of goods set off for Glasgow on weekday mornings.

In order to head off the threat to the town's livelihood, the influential men of Bo'ness agreed to make a branch from the Forth & Clyde Canal at Grangemouth to Bo'ness harbour, for which purpose the Borrowstounness Canal Company was formed and shares were offered for sale.

A contemporary account by one Mr M'Kenzie said,

Two Acts of Parliament and subscriptions to the amount of £10,000 were obtained. The canal was cut from the river Avon eastward within a mile of the town, and an aqueduct across the Avon was nearly completed; but after an outlay of about £7,500 the work was abandoned when not half-finished. The circumstances which prevented the accomplishment of this desirable undertaking … are, deeply regretted by the inhabitants of this town, especially on seeing their trade turned into another channel. Much of it passed by the canal direct to Glasgow, and the larger vessels discharged at Grangemouth, which was only a creek of [Bo'ness], but then became its rival, and was eventually erected into a separate port.

The promoters of the Bo'ness Canal were no doubt delighted that subscriptions went well at first, on the basis of which work was begun, initially according to a survey carried out by a Mr Lowrie. After spending more than £7,000 of the capital subscribed, more funds were needed as construction of the canal had turned out to be more difficult than anticipated. Construction therefore came to a halt.

An account by a Mr Johnston said:

> No doubt want of funds was one of [the reasons] but, if local tradition speaketh truth, a portion of the money raised was not expended as it ought to have been, and some associated with the project rendered themselves richer in pocket and poorer in character by their conduct at that time.

Nevertheless, hopes of resuming work were still held out. The committee resolved to seek another survey and a report on the canal from a more skilled engineer than Mr Lowrie, as well as an estimate of the cost of completing the canal, so that the necessary funds might be raised. Robert Whitworth, who oversaw the Forth & Clyde Canal, was employed to take the project forward.

Whitworth's report and estimate, a long one, dated 'Glasgow, 28 December 1789', was addressed to the proprietors, and began:

> Gentlemen, In obedience to your orders, I have taken the levels and made a survey of the line of the proposed canal from Grangemouth to Borrowstounness, and made a plan and profile of the same. The line laid down … is nearly the same as that laid down by Mr John Lowrie … The level of the reach of the Great Canal (the Forth and Clyde) above the second lock suits the level of the country very well, as appears by the profile, and from Grangemouth to the river Avon will be exceedingly easy to execute, except at the crossing of the Grange burn, where a small aqueduct will be necessary.

Beginning at Bo'ness, Whitworth's plan showed the harbour and then, heading west, there was a large canal basin. Somewhere in the vicinity of what was formerly the West End Foundry, there was to be a supply pond. Some way further on was Kinneil Castle. From Parknook, the canal veered away from the road and ran along the side of the Firth until it neared the mouth of the Avon. It then headed left and, from there, ran to somewhere near the present bridge over the Avon, which is also where an aqueduct was partially constructed.

On the plan, there is a strong dotted line running from Parknook along the side of the present road, and then changing direction at the Haining towards the Avon, almost on the same line as the present Grangemouth Road. This was evidently thought to be a more direct, and less expensive, route than the old one along the shore and through the fields of the Lowries. Whitworth was asked to survey the route indicated by the dotted line, and report on what he thought of adopting it. He duly reported that while he considered the line was perfectly feasible, as the shore route was on the way to being completed (and that a large amount of money had already been spent on it), he would not advise that it should be abandoned.

Whitworth's report continued,

> From the Avon aqueduct to Parknook the canal has been finished for seven feet of water; but as it has now to be raised to eight feet, the puddle will want raising on both sides the whole way. The south bank is rather slender, it having been formed with great economy for only seven feet, yet has stood so long to consolidate that it may sustain eight feet without enlarging the base.

He also reports in minute detail on the various sources which would require to be tapped along the route to supply the canal with water, but he was also relying on a supply from the Forth & Clyde Canal.

The estimate was based on the assumption that the canal would be 54 feet (16.5 metres) wide at the waterline and 27 feet (8.2 metres) at the bottom, with a depth of 8.5 feet (2.5 metres) as otherwise it would be too shallow for a vessel drawing 8 feet of water.

The total estimated cost of the canal from Grangemouth to the West Engine was £10,406 7s 6d, and from there to the harbour, including a proposed lock near the harbour and other items connected with the terminus, another £7,357 2s 6d, making a total of £17,763 10s.

The report and estimate meant that nearly £20,000 would have to be found – double what had been raised originally – which was utterly beyond the means of even the most enthusiastic of the promoters.

For the next twenty years, Bo'ness was fairly prosperous, with coal shipments, shipbuilding and other industries. However, on 1 December 1810, a customs house was opened at Grangemouth, which had previously only been an outstation of Bo'ness. In that year, the total duties levied at Bo'ness amounted to £30,485 17s. Five years later the figure had declined to £3,835 6s 4d. As a result, the number of skippers and sailors who lived in and around Bo'ness fell dramatically. Similarly, a great many well-known and substantial ship owners and merchants left.

The rise of Grangemouth eventually eclipsed Bo'ness, although at the end of the eighteenth century the town still had twenty-five ships. As well as shipping, the Bo'ness area had a number of industries that included coal mining, salt-making, shipbuilding, pottery manufacture and iron founding.

Ordnance Survey Grid Reference: NS 92612 80941

Campbeltown & Machrihanish Coal Canal

Near Stewarton, traces can still be found of the embankment for the Campbeltown Coal Canal. The canal was 3 miles (5 kilometres) long and was completed in 1794 to carry coal mined at Drumlemble.

Coal had been mined on the Kintyre peninsula since at least 1498. Although it was not of the highest quality, the coal seams were plentiful and relatively cheap to mine. In the middle of the eighteenth century, the collieries had customers close by in the many whisky distilleries of the area. In 1773, James Watt was asked to survey a route for a canal from Drumlemble to Campbeltown in order to reduce the costs of transportation. Building started in 1773 and the canal opened in 1791.

Once again, for local people and industry to invest in a canal is an indication of the difficulties of transport by road, as well as the hazards of going by sea. Campbeltown is one of Scotland's most remote mainland towns. It is 89 miles (143.2 kilometres) south of Oban and 134 miles (215.6 kilometres), by road, from Glasgow. It has had various industries over the years, including coal mining, shipbuilding and fishing: in the late nineteenth century there were reported to be 646 boats in the harbour. There were thirty-four distilleries and many people were, until recently, employed in the manufacture of clothing.

Evidence of the canal's existence can still be seen between Hillside and Gortan Farms, and there are also the remains of the aqueduct which crossed the Chiscan Water.

From north of Gortan, the canal turned south towards North Moy and South Moy. Beyond the latter it turned west, passing through the farms of Knockrioch and Tonrioch, before crossing the Chiscan Water near the present road bridge on the Machrihanish road. From Lintmill, the canal continued westwards, and its dry bed can be seen running parallel to the later light railway track north of Dalivaddy farmhouse.

The canal's course is evident in a number of places. A ploughed-down embankment is still prominent enough to indicate the raised course of the canal, after which a broad depression, disrupted by later quarrying, marks a former cutting. The next length of the canal bed has been infilled and is under pasture. After the Chiscan Water, there is the best-preserved section of the canal: although it has been partially infilled and the banks are broken, it is 1 foot 6 inches (0.5 metre) deep. The light railway was built on the towpath of the canal.

The canal became disused and was virtually abandoned by 1856. When the Argyll Coal & Canal Company acquired the main colliery in 1875, the canal was found to be in a poor state of repair. The company decided that a better means of transport was needed, and promptly began to plan the building of the light railway to Campbeltown.

Ordnance Survey Grid Reference: NR 69631 19831

Carlingwark Canals

The Carlingwark Canal, on the lower reaches of the River Dee near Kirkcudbright, was one of the earliest Scottish canals to be completed. In 1765, a 1.5-mile (2.4-kilometre) cut was made from Carlingwark Loch to the Dee near Threave Castle. Known in the area as the Carlingwark 'Lane', a local word for a slow flowing stream, the canal cut a straight course across the Carlingwark and Blackpark mosses or bogs.

It was the work of Sir Alexander Gordon of Culvennan, and reduced the level of the loch by several feet to enable marl to be dug out of its bed, and was used to transport the marl to neighbouring farms for use as a fertiliser.

The Dee was improved in around 1780 by the construction of a second cut, 0.5 mile (0.8 kilometre) long, near Culvennan House, which bypassed a fast-running stretch of the river. There was one, or possibly two, locks near the house. The canal was used by barges which carried their loads of marl upstream to farms up to 15 miles (24.1 kilometres) along the River Dee.

Some years later, a far more ambitious scheme was proposed for a 26-mile (41.8-kilometre) canal with a number of locks. The Glenkens Canal would have run from Dalry, alongside the rivers Ken and Dee, to the estuary near Kirkcudbright. Surveyed by John Rennie, it was authorised by an Act of Parliament in 1802. Rennie estimated that it would cost more than £13,000. However, it did not receive enough financial support. There are, though, visible remains, indicating that some work was done on the ground, just south of the Glenlochar barrage and bridge.

Marl-carrying boats certainly used the Carlingwark Canal until about 1840 but, as it was undertaken privately by Sir Alexander Gordon, there was no need for an official abandonment. It is therefore impossible to say whether it was used at all after 1840. Both lengths are still extant. It was still shown on the 2007 Ordnance Survey maps, where the local term 'Carlingwark Lane' is used.

Ordnance Survey Grid Reference: NX 76332 61263

Carlingwark Canal Pumping Station. Image used by permission of David Bain.

Carron Canals

The Royal Commission on the Ancient and Historical Monuments of Scotland (RCAHMS) dates the construction of the first Carron Canal to 1781. This was concluded as a result of the discovery of an inscription uncovered during an archaeological dig in the 1990s. The canal ran east from the Carron ironworks, parallel to the River Carron for a distance of about 0.5 miles (804.6 metres), where it passed through a lock to join the river. This was at the highest navigable point on the river at that time. The date of the canal's eventual abandonment is not known but it is shown on maps dating to 1864.

The Carron Company had been established in 1759 on the banks of the River Carron near Falkirk, Stirlingshire. It was founded by two Englishmen, Dr John Roebuck, a chemist, and Samuel Garbett, a merchant, together with a wealthy Scottish shipowner, William Caddell, whose son was appointed manager. After initial problems, the company was at the forefront of the Industrial Revolution. It prospered through the development and production of a new short-range, short-barrelled naval cannon, the carronade. The company was one of the largest ironworks in Europe throughout the nineteenth century. It later became one of several foundries producing pillar boxes and was one of five foundries that cast Sir Giles Gilbert Scott's classic red telephone boxes. After 223 years, the company became insolvent in 1982 and was acquired by the Franke Corporation, rebranded as Carron Phoenix, and it is still operating today.

The second canal, the Carron Cut, was also built by the ironworks in the second half of the nineteenth century. It was also about half a mile long, joining the River Carron to the Forth & Clyde Canal at the latter's Lock 3. It created a short-cut from the Forth & Clyde Canal to the ironworks, thus avoiding a journey down to Grangemouth docks and back up the river. A small section at the northern end can clearly be seen on aerial photographs, and the rest of its line can be clearly identified.

Both the Carron canals were built to transport iron ore and coal to the ironworks, and to export the finished products. When the Forth & Clyde Canal was reopened in 2001, a new cut was made slightly to the east of the original alignment because, by then, the section to Grangemouth had been heavily built on. This new section was given the name the 'Carron Cut'.

Ordnance Survey Grid Reference: NS 89701 82718

Fleet Canal

Gatehouse of Fleet is a village situated near the mouth of the Water of Fleet, in Dumfries and Galloway, ten miles west of Kirkcudbright. It lies at the head of Fleet Bay, where the Water of Fleet joins the Irish Sea.

In the late eighteenth and early nineteenth centuries, the village was a thriving industrial centre with cotton mills, shipbuilding, a brewery and its own port. It was known locally as the 'Glasgow of the South'.

Gatehouse of Fleet was some distance up the River Fleet and, as boats became larger, it became more difficult to get them close to the town. In 1824, the twenty-one-year-old Alexander Murray, of the Stanhoe Murray family, whose estate was at Broughton in south-west County Donegal, decided to dig a canal to the sea to bypass the obstacles in the river. He brought over men who were in arrears of rent from his estates in Donegal to carry out the work.

The canal was 1,440 yards (1.3 kilometres) long and ended at Port McAdam, situated on Fleet Street, where a wharf was built that was in use up to the 1930s. The line of the canal is shown on current OS maps, although it is marked as the Water of Fleet.

Ordnance Survey Grid Reference: NX 59679 55985

Forth & Cart Junction Canal

The idea of a direct connection between the Cart Navigation and the Forth & Clyde Canal was first suggested by Hugh Baird in 1799, but no action was taken at that time. When the Forth & Cart Junction Canal was promoted in the 1830s, it effectively revived Baird's plan. The intention was to provide a better route between Paisley and the Firth of Forth than the alternative – which went by way of Port Eglinton and Port Dundas.

The Junction Canal was very short, just 0.5 mile (0.8 kilometre) long, but it provided a short-cut. Its purpose was to link the town of Paisley, the Firth of Forth and Port Dundas, Glasgow, without the need for vessels to go via Bowling on the Clyde, about 7 miles (11 kilometres) downstream.

An Act of Parliament was obtained in 1753, authorising improvements to the River Cart to make it navigable and, by doing so, to assist the growth of the cotton industry in Paisley. The works included making the channel of the Cart straighter and deeper.

When a new turnpike road bridge was built in 1787 at Inchinnan, the Paisley Town Council obtained a second Act of Parliament, authorising the construction of a new navigable cut which would pass under the turnpike road. Work started on 23 August 1787, and was completed within a year. With the anticipated building of the Junction Canal, further work was carried out in 1835 to improve the harbour facilities at Paisley.

It was also expected that coal from Coatbridge would reach Paisley through the Monkland Canal, the Forth & Clyde, the Junction Canal and the Cart Navigation. It would also have the advantage of saving water on the Forth & Clyde, as smaller boats could pass through the link, rather than using the larger locks to the west, where the Forth & Clyde joined the River Clyde.

The Forth & Cart Junction Canal was authorised by an Act of Parliament in May 1836 and was completed in 1840. It had three locks; a single lock and a pair of staircase locks, each 67 feet by 15 feet (20 metres by 4.5 metres) long. The total rise was 30 feet (9 metres).

The Forth & Clyde Canal obtained an Act of Parliament in 1842 to enable it to take over the Forth & Cart, but this did not happen for another thirteen years, by which time the revenue of the Junction Canal was only £325 per annum, while the company had expenditure on maintenance of the canal, and was still paying interest which was nearly equal to its income on its debts from the time of its construction. The canal was in poor condition, and it was estimated that it would cost £3,100 to return it to good order. The annual traffic was by then down to about 40,000 tons (40,614.8 tonnes) as the coal that Paisley used was now supplied by railway. The Forth & Clyde estimated that they made £739 per year from traffic that passed from the Junction Canal to their canal, so they offered to buy the Junction Canal for £6,400. If traffic was greater than 90,000 tons (91,444.2 tonnes), the original proprietors of the Junction Canal would receive an extra penny per ton.

Although the payment would not clear the debts, and the likelihood of traffic increasing sufficiently to generate the payments was slight, the committee had little option but to accept. The Junction Canal was taken over by the Forth & Clyde Company in 1855. In its turn, together with the Forth & Clyde Canal and the Monkland Canal, the Junction Canal was taken over by the Caledonian Railway in 1867.

As it was unprofitable, the Junction Canal was closed in 1893, providing an opportunity for the Glasgow–Clydebank railway to extend its line to Dalmuir. The line opened in 1896, destroying the first section of the canal. By 1937, the north end had been built on, leaving no trace of the canal. At its south end, the canal was cut short just below the first lock, where the Lanarkshire & Dunbartonshire Railway crossed it. It remained like this in 1919 but, twenty years later, it had been reduced by half again. A small part of it was still visible in 1985, but now no trace remains.

Ordnance Survey Grid Reference: NS 48026 68925

Glasgow, Paisley & Ardrossan Canal

The Glasgow, Paisley & Ardrossan Canal, which was later known as the Glasgow, Paisley & Johnstone Canal, ran between Glasgow and Johnstone. As designed, it would have been just under 33 miles (53 kilometres) long. The canal would be 30 feet (9.1 metres) wide at the waterline and 18 feet (5.5 metres) at the bottom, with a depth of 4 feet 6 inches (1.3 metres). Although the original intention had been to build the canal as far as Ardrossan, it never reached the town. It ran from Port Eglinton in Glasgow to Thorn Brae in Johnstone. The canal was the scene of a major human tragedy within just a few months of its opening.

The canal had first been proposed by the 12th Earl of Eglinton in 1791, with the aim of connecting the growing industrial centres of Glasgow, Paisley and Johnstone with the earl's newly built deep sea port, on which he had spent the colossal sum of £100,000, and with his coal fields in Ayrshire. He was intent on making Ardrossan the main port for Glasgow in order to recoup his investment. The harbour was eventually competed by his son, the 13th Earl of Eglinton, at a total cost of £200,000.

Eglinton's fellow shareholders in the Glasgow, Paisley & Ardrossan Canal Company included another colliery owner, William Dixon of Govan, who also wanted to find new markets for his coal. Interest in the canal was also expressed by Lord Montgomerie and William Houston, both of whom would benefit from it passing through their lands, and connecting their collieries and iron mines to a greater number of industrial users. In the period before McAdam invented his all-weather surface, the roads in Lanarkshire, Renfrewshire and Ayrshire were simply not suitable for heavy traffic. The only other alternative was to use the Clyde estuary to Glasgow, but it was too shallow to be navigated by large vessels.

This was not to be a project that failed for lack of investment in the best professional advice: the engineers John Rennie, Thomas Telford and John Ainslie were all employed to survey a route and to provide an estimated cost. The three engineers' design was for a contour canal about 11 miles (18 kilometres) long with no locks, although it would mean that the total length of the canal would be longer. After that there would be a series of eight locks to lift the canal to its summit near Johnstone. Finally another thirteen locks would bring it down to sea level at Ardrossan Harbour.

An Act of Parliament, which received the royal assent on 20 June 1806, incorporated the Glasgow, Paisley & Ardrossan Canal Company, with Eglinton as the chairman. It permitted capital to be raised by the sale of 2,800 £50 shares (£140,000 in total) of which the proprietors, the Earl of Eglinton and Lord and Lady Montgomerie and Lady Jane Montgomerie, subscribed £30,000.

Construction began in 1807 between Glasgow and Johnstone, with Telford himself apparently acting as engineer, and it progressed without the contractors encountering any great difficulties. Telford also designed the aqueduct that carried the canal over the

White Cart River, at Blackhall in Paisley. The canal was opened to Paisley, including two short tunnels with towpaths, in 1810, and to Tradeston on the south bank of the Clyde in Glasgow in 1811. The terminus at Tradeston was named Port Eglinton. Having got that far, the company ran short of funds and £300,000 of additional funding was needed to complete the project. The proprietors asked the government for the money to continue the canal to Ardrossan. However, improvements had been made to the Clyde, and there was no need for the canal to go as far as Ardrossan, so the government refused its help.

Even so, the 11-mile-long (17.7-kilometre) section of the canal that had already been opened saw a busy and growing traffic. Stone, manure and soil were charged at 2*d* per ton per mile; coal, coke, culm and lime were 3*d* per ton; bricks, tiles, slates, iron and metal were rated at 5*d* per ton; and all other goods were charged 2*d* per ton. In 1840, the canal handled 76,000 tons (77,000 tonnes) of cargo. There was ferocious competition between the canal and road carriers. The cotton cargoes from the mills at Johnstone were carried by specially designed fast boats, similar to narrowboats on English canals.

A swift passenger packet service was started. The first boat, *The Countess of Eglinton*, had been launched on 31 October 1810. Long and shallow-drafted, the wrought-iron boats carried sixty passengers at an average speed of 8 mph (13 kph) including intermediate stops. The hulls were constructed of light iron ribs clad with thin wrought-iron plates, while the cabins were covered with waterproof oiled cloth. The boats had a maximum speed of 10 mph (16 kph) and made fourteen journeys a day. They were towed by teams of two horses which were changed every 4 miles (6.4 kilometres). Short-haul traffic was carried by so-called 'market boats', and there was a special 'Glasgow Packet'.

The idea of the fast passenger boats was widely copied by the owners of other canals, and this type of boat were became known as 'swift boats' or 'fly-boats'. An inn was built at Port Eglinton in 1816, and a wharf was built on the north bank of the White Cart near Crookston Castle. Canal basins were also built at Paisley and Johnstone. By the 1830s, the journey from Glasgow to Paisley took just one hour eleven minutes.

The Paisley canal disaster, one of the worst on any canal in Britain, cost the lives of fifty-two men and thirty-three women on Saturday 10 November 1810, the day of the Martinmas Fair and only four days after the canal had been opened. Many people were given the day off work, and took the chance to travel the short distance between Paisley and Johnstone on the canal. As *The Countess of Eglinton* docked at the Paisley wharf, people rushed to get on the boat while those from Johnstone tried to disembark. The boatmen tried to push off again, but the weight of people caused the boat to overturn, and many people were thrown into the water. Despite the shallowness of the water, its coldness and the sheer side of the wharf made it impossible to climb out, and the problem was compounded by the heavy winter clothes people were wearing, and the fact that few town dwellers at that time could swim.

The canal was bought in 1869 by the Glasgow & South Western Railway Company and, in 1881, it was abandoned by Act of Parliament. 200 years after the disaster, a plaque was unveiled by the then Provost of Paisley, Celia Lawson, to commemorate the eighty-five who drowned in the canal.

Ordnance Survey Grid Reference: NS 60282 64122

Inverarnan Canal

The Inverarnan Canal was originally built to provide a sheltered berth for the steamboats that operated on Loch Lomond in the 1830s and had become the most popular means of travel on the loch. The steamers were used by local people and, more importantly, by growing numbers of tourists, for whom it was the preferred means of transport from Clydebank to the Highlands.

Originally the steamers moored at Ardlui, where passengers and their luggage and goods from Glasgow were transferred to the shore by rowing boats to meet the waiting stagecoach. In certain weather conditions, this transfer was difficult if not impossible.

In 1840, the possibility of making the lower part of the River Falloch navigable for this purpose was investigated. The idea was suggested and supported by the owners of the Loch Lomond Steamer Transport Company, Mr Napier and Mr McMurrick.

The project began after expert opinion (whose is not known) had recommended that, if a number of bends in the River Falloch could be bypassed by a canal leading from the river to Inverarnan, it would be feasible for the steamers to offer a more reliable service.

The anonymous expert also pointed out that, in drought conditions, the river would need to be dredged in order to ensure a sufficient depth of water. As the local heritage group points out, 'Loch Lomond is not usually associated with deficiencies in water', so, on reflection, Messrs Napier and McMurrick decided that a cheaper and simpler solution would be to build a jetty at Ardlui. Even this was not done until another ten years had elapsed. Although it is known that there was an existing pier on the west bank of the River Falloch at Ardlui, there is no evidence that the steamboat company used it.

The construction of the canal and its associated works finally began in 1842. The man responsible for supervising the work was a Mr Ferrier, the father-in-law of the company's steamboat manager, Thomas Mclean. Construction was delayed by a combination of wet weather, heavy frosts and snowfalls. It was finally completed, however, in 1844.

The canal, as documented in the steamboat company's archives, was 530 yards (484.6 metres) long and terminated at a turning basin with a landing stage about 290 yards (265.1 metres) south of the 'Inverarnan Inn'.

Before 1842, there is no specific mention in the archive of Inverarnan or the canal. It was generally referred to vaguely as the 'Head of Loch Lomond'. Even in 1843, the only mention is of the Loch Lomond steamer sailing from Balloch at 10.00 a.m. for the head of Loch Lomond.

It is recorded that the *Loch Lomond* was the first steamer to navigate the River Falloch and the Inverarnan canal. Evidently Messrs Napier and McMurrick were entrepreneurs to their finger tips. They publicised the canal, advertising it as the highlight of a trip on their steamers.

Both passengers and goods were carried, and the coaches to Killin and Ballahulish now started from the 'Inverarnan Inn', and later services from the inn to Oban, Inverness (by way of the Caledonian Canal), Aberfeldy and Crieff were added.

By the mid-1850s, other operators and businesses realised the usefulness of the direct link between the loch and coaches. In May 1858, Mr Andrew Menzies, on behalf of the Glenorchy & Glencoe Coach Company, requested that the steamboat company allow the boats to go up the River Falloch at all times, weather permitting. At first, this request was ignored, resulting ultimately in the abandonment of the canal. Eventually, however, the company instructed their captains to take their vessels up the River Falloch whenever it was practical to do so.

In November 1858, Thomas Mclean, the steamboat manager, was instructed to ascertain whether the service using the River Falloch had met with public approval. Possibly significantly, Mclean was asked if a Captain Brown had taken over the Ardlui Pier lease, and if personal interest was influencing the company's business and the destination of its steamers. The manager's reply regarding the reactions of customers is not known, but it is documented that he found Captain Brown had indeed leased the Ardlui Pier, although he had promised to give it up at the end of that year.

Mclean also found that the directive to the steamers' captains, to take their vessels up the River Falloch, had been ignored. A number of inn-keepers and farmers wrote to the steamboat company claiming that the steamers did not go up the river, even when there was no obvious reason for not doing so. All the steamers certainly called at Ardlui, but whether they then continued up the River Falloch to Inverarnan, the 'new' Garabal Landing or the old Garabal Landing seemed to depend on the whim of the captain, and this was affecting local businesses. This time, the captains were told to take the steamers up the Falloch, provided there was a sufficient depth of water to do so. To avoid any argument, the company installed a water gauge at Ardlui.

In 1860, the steamers were carrying mails to and from Luss, Rowardennan, Tarbet, Inversnaid and Ardlui, but apparently not to Inverarnan, although it continued to feature in coach advertisements during the 1860s. This suggests that the canal was still being used until at least 1868. However, from 1870 coaches were advertised as running from the head of Loch Lomond. Although the term had previously been synonymous with access by the canal to Inverarnan, it was now being used for the general area of Ardlui.

The opening of the Callander & Oban Railway as far as Killin (Glenogle) on 1 June 1870 sounded the death knell for through services to the north going via Loch Lomond. It was now quicker, more comfortable and more convenient to travel by train to Killin and to join northbound coaches there. Navigation of the canal and the River Falloch simply faded away, although a trip on the passenger boats on the loch is still a popular excursion for tourists.

Although the canal still exists, overgrown and unused, an adventurous canoeist explored it in his outrigger sailing canoe to see how far it was possible to navigate the River Falloch. It was autumn, and the absence of leaves on the trees allowed him to easily see the entrance to the canal, which he had not found previously. Trees overhang the canal along the whole of its 530-yard (450-metre) length and when the branches are covered in their summer growth of leaves, the canal is almost impossible to find. He succeeded in paddling all the way up to the turning basin at the northern end of the canal.

Ordnance Survey Grid Reference: NN 32120 18178

Kilbagie Canal

The Kilbagie Canal was one of the first canals in Scotland in the canal era. It was a little over 1 mile (1.6 kilometres) long with an aqueduct, and ran between Kilbagie and the port of Kennetpans. It opened in 1780 and was in use until 1861, when it seems to have been replaced by a tramroad. The canal, which was south-east of Alloa, Clackmannanshire, ran to Kilbagie from the wharf at Kennetpans on the River Forth.

A distillery had originally been founded at Kilbagie by John Stein in around 1720, and whisky was produced there until 1845. After that date, the Kilbagie Chemical Manure Company used the distillery site to produce chemical fertiliser until the late 1860s. The buildings were then used by J. A Weir for the production of fine papers using esparto grass pulp.

Kennetpans was built largely on marshland that had been reclaimed from the River Forth by depositing coal ash from the salt pans, which gave Kincardine its first industry and original name of West Pans. By the middle of the sixteenth century, salt manufacturing had become firmly established in the area and there were outcrops of easily mined coal nearby that fuelled the process of salt making by evaporation.

A surviving row of miners' cottages at Kennetpans, which are listed buildings, were built by Lord Dundonald. One of the other reasons for the success of the coal pits at Kennetpans was that the coal could also be exported from its small port. The other export, of course, was whisky from the Stein's Kilbagie Distillery.

Ordnance Survey Grid Reference: NS 93161 90066

Monkland Canal

Once the most successful of the Scottish canals, the Monkland Canal was built primarily to supply Glasgow with cheap coal. It opened in 1793 from the River Calder near Woodhall, south of Airdrie, to a basin not far from Glasgow Cathedral.

The canal was projected principally to bring coal from the Lanarkshire coalfield as early as 1769, when James Watt made a survey. The project was well supported by investors, the canal's Bill was passed by Parliament in 1770 and Watt became the engineer, although he resigned in 1774. By 1773, when the canal had reached Barlinnie prison north-west of Glasgow, it stopped while further capital was raised, and it did not restart until 1782, when cutting recommenced westward to Blackhill (where four staircase locks were later built). From there, the cut continued at a lower level into the centre of Glasgow.

The Blackhall locks would require a greater supply of water, so another Act was passed in 1790 that authorised an extension eastward to the River Calder at Woodhall. Thereafter, the canal was in effect a feeder to the Forth & Clyde, which it was to join at Port Dundas. By late 1790, the Monkland Canal was complete for 12.25 miles (19.7 kilometres) from Woodhall to its basin in Glasgow, with two locks at Sheepford and four staircase pairs at Blackhall, all of which could accommodate scows measuring 66 feet by 13 feet 6 inches (20.1 metres by 4.1 metres). Traffic developed steadily as the north Lanarkshire collieries became more productive.

The canal eventually had four branches. The earliest, 1 mile (1.6 kilometres) long, was built at the expense of the Calder Coal Company in 1799/1800 and was called Dixon's Cut. The Gartsherrie Branch, less than half a mile (0.8 kilometre) long, was cut in 1827 to serve two pits and then the Gartsherrie ironworks, which was built in 1828. The quarter-mile (0.4-kilometre) Langloan Branch served a colliery and, after 1841, another ironworks. Finally, the quarter-mile (0.4-kilometre) Dundyvan Branch, built after the canal was acquired by a railway, served the coaling basins at Palacecraig and its railway owners.

Railway competition had arrived as early as 1826 with the completion of a horse-drawn tramroad from the Monkland collieries to the Forth & Clyde Canal at Kirkintilloch. Five years later, the locomotive-hauled Glasgow & Garnkirk Railway was built, from Port Dundas to Gartsherrie, running more or less parallel to the canal.

Initially, neither line worried the Monkland Company too much. Passenger boats were even introduced, with the passengers changing boats at the Blackhall locks to save lockage time. The locks had to be doubled in 1841 for the heavy coal traffic. By that time the Monkland Company had experimented with a steam tug, although it was sold in 1846, the year in which the Monkland and the Forth & Clyde Canals amalgamated.

The amount of coal traffic was placing a strain on the water supply, but this was solved in 1850 by the opening of an inclined plane at Blackhall for returning empty boats to the pits, and a reservoir that was built in 1852. During the 1860s, steam scows were beginning to replace the horse-drawn ones, although they were limited in size by the fixed bridges eastward from Blackhall.

However, even with these innovations, the canal found itself assailed by competition from the Caledonian Railway, and by the new rail-served coalfield that was developed around Wishaw. The final blow was in 1870, when most of the Monkland's traffic was captured by the new North British Railway between Glasgow and Coatbridge.

The Blackhall-inclined plane was closed in 1887 for want of traffic. By the first decade of the twentieth century, little freight was moving on the canal and the iron trade was finished completely. All traffic ended by 1935, but the canal was still needed as a feeder to the Forth & Clyde and a power station, and there was no question of abandonment unless the owners – by that time the London, Midland & Scottish Railway (LMS) – agreed to provide piping for the water. In spite of the parlous state of the canal, and frequent drownings in the Blackhall locks, work on infilling and piping the Monkland was delayed, even after the Act of Abandonment was passed.

The canal was legally abandoned for navigation in 1942, but it was not until 1954 that infilling began. Today only fragments of it remain; some sections were culverted and built over, and most of it is now buried under the M8 motorway. Even so, three sections remain in water, where local initiatives have breathed new life into the canal. While recognising the Monkland's important place as part of Scotland's industrial heritage, it has been transformed into a community green space.

Ordnance Survey Grid Reference: NS 34490 73945

Monkland Canal. Image used by permission of John Alexander Rae.

Monkland Canal, Bargeddie. Image used by permission of Brian Cairns.

Muirkirk Canal

The Muirkirk Canal in Renfrewshire, 1 mile (1.6 kilometres) long, was built to serve the needs of an ironworks that was established in 1789, in what was described as the 'desert and inland' parish of Muirkirk. The canal was built to take advantage of available and cheap coal, ore and limestone from a recently discovered bed of limestone at Ashieburn, which became known as Newhouse Quarry. The canal was opened in the same year as the ironworks.

According to the records of the ironworks managers, on 27 September 1790 they resolved 'to make a canal 8 feet (2.4 metres) wide at the bottom, 16 or 17 feet (4.8 or 4.95 metres) wide at the top, and 4 feet (1.2 metres) deep'. Permission was to be sought to bring a supply of water through lands owned by Mr J. L. MacAdam, of Craigengullen, at Ashyburn. Cargoes included coal from the Lightshaw, Auldhouseburn and Crossflat collieries, which were connected to the canal by 'bogie' – or tramroads. The cargoes were carried on raft-type barges.

The remains of the Muirkirk Canal can be followed over a distance of some 1.5 miles (2.5 kilometres) from the Ashaw Burn to the site of the ironworks. The canal served two purposes: supplying the ironworks with coal and limestone from the eastern part of the Renfrewshire mineral field, and providing the works with water power to operate the blowing engine and a forge.

The canal's original water supply came from the River Ayr but, as industry grew in the area, including mills at Catrine, two reservoirs were built upstream on the river to ensure that there would be a sufficient head of water. The upper reservoir, known as Glenbuck Loch, some 3.75 miles (6 kilometres) to the east of the ironworks, was created in 1802 and is still used to store water. However, the lower reservoir, built in 1808 a short distance downstream, was deliberately breached at some time in the past and the only visible sign of its existence is a grass-covered embankment.

There was a substantial embankment on the north side of the canal, wide enough to accommodate a towpath along much of its route. Its course was dictated by the gradient, and aqueducts had to be built to carry it across three streams. The terminal basin at Muirkirk was to the south-east of the furnace bank at the ironworks, and an outflow to the north supplied the water needed to power the blowing engine at the ironworks.

With the closure of the ironworks and its subsequent demolition in the late 1960s, the canal basin was infilled, but the course of the canal, although it is de-watered, can still be traced running to the east. Although the first edition OS map shows the canal, by 1896 when the second edition OS map was published, it had been superseded by a mineral railway and was no longer in use. The remains of the canal can be seen by following the route of the former Auchinleck to Muirkirk branch line. The walk follows the routes of the old railway line and the canal, with part of it following the canal bank.

Ordnance Survey Grid Reference: NS 69620 27325

Pudzeoch Canal

Beside the landing stage for the ferry at Renfrew is a small tidal basin, known as the Pudzeoch Basin, which until recent years was filled with a variety of small craft owned by Offshore Workboats Limited. What is not widely known is that it is a section of an aborted canal, half a mile (0.8 kilometre) long, to run to Ayrshire, and which ran beside Canal Street from the basin. Exactly when the canal, which was the canalised Pudzeoch Burn, was built is not known, as there is no record of its original date, nor of who planned it, in any archive records. Local folklore says that the canal was intended to connect with the Glasgow, Paisley & Ardrossan Canal. It is also said that it used to run a lot further inland than it does today but, again, this cannot be confirmed.

However, its existence is confirmed by nineteenth-century maps, and by a Renfrew Council report on plans to develop the river bank which refers to 'consolidating the ground beside the Pudzeoch Canal'. Part of the canal had already been infilled by 1924, and by 1939 all of it except the basin had been infilled, although, by then, the basin itself had been widened.

Ordnance Survey Grid Reference: NS 50812 67705

Sir Andrew Wood's Canal

Sir Andrew Wood, Scotland's greatest admiral in the reign of James IV of Scotland, is often referred to as the Scottish Nelson. He had made his name in battles against the English fleet in the years around 1500 and, in recognition of his victories, he was given estates in Fife by James IV. The connection with Sir Andrew makes the canal that he built by far the earliest canal to be built in Scotland.

Sir Andrew lived at Largo House, 300 yards (274.3 metres) west of the parish church at Upper Largo (Kirkton of Largo) in Fife. Only an early seventeenth-century tower, known as Wood's Tower, now survives of the house. Sir Andrew linked his house and the church with the canal in about 1495, simply so that he could be rowed to church in a manner he thought was fitting for a naval hero.

Remarkably, the remains of the canal can still be traced in the field behind Largo Home Farm. Its course runs parallel to, and on the north side of, an old, ornamental enclosure bank. Other remains are evident as a slight scarp and a short stretch of a shallow channel. Elsewhere there is no trace of the canal.

As the area is dominated, geologically, by rocks of the Devonian and Carboniferous periods, underlaid with sedimentary rocks such as old red sandstone and igneous rocks, it may have been the case that the canal could have held water without puddling. Certainly, excavations by archaeologists in recent years have revealed no traces of clay puddling.

Built with no locks or bridges, the canal was abandoned some time before 1790. Some people are of the opinion that, if anything, it was a drainage ditch. However, according to the Upper Largo Conservation Area Appraisal of March 2012, there is archaeological evidence of the canal, approximately 0.3 mile (0.5 kilometre) in length. The feature was marked as 'Canal (remains of)' on the Ordnance Survey (OS) 1:10,000 map of 1987, and on the first edition of the OS 6-inch map, sheet 25, Fife, of 1855, there was a line which followed the same course.

Ordnance Survey Map Reference: NO 42526 03483

St Fergus & River Ugie (Pitfour's) Canal

Documentary sources of any kind about the St Fergus & River Ugie Canal are extremely hard to find, but one man, Angus Graham, has followed the remains of the canal on the ground and recorded them in detail.

The idea of cutting a canal to open up the country inland from Peterhead was being discussed as early as 1793, and before the end of the century work had been started by James Ferguson of Pitfour, who built a 4-mile (6.4-kilometre) canal in St Fergus parish, primarily for the benefit of his estates there and in the Old Deer and Longside parishes. Ferguson had originally proposed a canal about 10 miles (16.1 kilometres) long, following the course of the River Ugie.

However, the work was never finished, owing to what Ferguson described as 'difficulties in effecting the necessary arrangements with neighbouring heritors'. Whether it was ever used for anything apart from supplying water is not clear, but it was certainly out of use by 1868.

The failure to reach Peterhead did not discourage Ferguson's ambition because, earlier in the nineteenth century, he had a branch of the canal cut northwards to Inverquinzie, near St Fergus, to carry calcium-rich shell-sand from the seashore, near the mouth of the River Blackwater, for use as manure. The sand must have originally been carried to Inverquinzie along a road which used to run past Scotstown.

Robertson and Gibb's map shows another canal crossing the flat lands near Scotstown. However, while this may have been Ferguson's intention, it would have required engineering of a higher order than the original canal and branch, as the flat lands are about 40 feet (12.1 metres) lower than Inverquinzie. In the event, the Inverquinzie branch likewise proved a failure; by 1837 it was regarded as useless except as a reservoir for water for the farms, and was already fast silting up.

The main canal runs along the north side of the valley of the River Ugie, which is formed by the confluence of the north and south Ugie Waters about 1.25 miles (2 kilometres) east-north-east of Longside. The *New Statistical Account* suggests that Ferguson may have intended to make one or both branches of the river navigable, but this, too, would have required major feats of construction since the Waters are small, shallow and twisting.

The St Fergus & Ugie Canal seems to have left the North Ugie Water at a point 130 yards (118.8 metres) north-west of the confluence but, today, there is nothing resembling an opening or sluice in the high left bank, nor is there any sign of a weir in the river. The first length of the canal has been made narrower and seems to have been converted into a feature reminiscent of a 'ha-ha'. However, there are some significant remains where the canal approaches the river bank some 500 yards (457.2 metres) downstream. North-west of the Haughs of Rora, 'Old Canal' is marked on the OS maps.

The canal crossed the Crooko Burn on a surviving aqueduct constructed with heavy granite blocks. The top of the aqueduct is dilapidated and overgrown, but the burn runs through a pair of openings with a cut-water. There is a third opening for a field ditch. It was clearly a substantial, well-designed structure.

About 120 yards (109.7 metres) east of the aqueduct, a small watercourse which runs more or less parallel to the Crooko Burn joins the canal. This was probably done, after the canal had been abandoned, to bring water to a small mill, some remains of which are visible on the north side of the line of the canal. A small watercourse runs under the canal in an original culvert, about 3 feet high by 2 feet 6 inches (0.9 by 0.7 metres) wide with an arched opening, at about the same point. The canal is better preserved here, no doubt because it was used as a reservoir for the mill, and perhaps also for a brick and tile works that used to stand nearby.

Continuing south-east, the canal has clay banks about 55 feet (16.7 metres) apart. In the next section, south of Cairnhall, the canal runs along the face of a slope which falls steeply to the Ugie. The channel here is about 20 feet (6 metres) wide by about 4 feet (1.2 metres) deep. The canal then bends away from the river and runs north-east to what looks like an overgrown pond next to Hallmoss Farm, which was the terminus of the Inverquinzie Branch. The section in what is now the garden of the farmhouse has been infilled, but beyond the farmhouse it is visible passing behind the abandoned Hallmoss Smithy. From the southern corner of the smithy, a much ploughed-down terrace continues south-east for some 130 yards (118.8 metres) but ends abruptly in the middle of a field, quite clearly at the point where building stopped.

The elevation above sea-level here is more than 60 feet (18.2 metres), but there is nothing to show how Ferguson intended to take the canal down to sea-level in the Ugie estuary or elsewhere. The high-water level in the estuary would have reached about 1,300 yards (1,188.7 metres) from the end of the canal. The alternative of trying to reach the Peterhead harbours, nearly 2.75 miles (4.4 kilometres) away, would have needed a substantial aqueduct spanning the Ugie.

The Inverquinzie Branch is just over 1.5 miles (2.4 kilometres) long. At its head the terminal basin marked on the 6-inch OS map of 1872 has now been infilled. From Inverquinzie, it followed a winding course to a point 160 yards (146.3 metres) west of Lunderton but, beyond this point, it has been infilled. The Cuttie Burn is crossed using another masonry culvert that is circular in section and 3 feet 6 inches (1 metre) in diameter. Where it is well preserved, the branch is about 18 feet wide by 5 feet deep (5.4 by 1.5 metres). Near Lunderton it remains partially in water.

Ordnance Survey Grid Reference: NK 12201 47460

Stevenston (Saltcoats) Coal Canal

In January 1772, work started on Scotland's first commercial canal in Stevenston, Ayrshire. It would be used mainly for carrying coal from neighbouring collieries to Saltcoats harbour for shipment to Ireland. The canal was the first of a number of Scottish canals that would be known as 'barge canals', built for a single user or small group of private users.

The Stevenston Coal Canal was just over 2 miles (3.2 kilometres) long with no locks, and was 12 feet wide and 4 feet deep (3.6 metres by 1.2 metres). Most of it was cut along the old course of the sea channel, which was a relic of the days when Ardeer was an island. Coal was carried on barges and the waste, instead of being disposed of on spoil heaps, was dumped along the route of the canal as a wind-break against blown sand, which was a frequent problem. The coal was carried from the seaward end of the canal to vessels waiting in the harbour at Saltcoats.

The canal was built for Robert Reid Cunningham of Seabank (now called Auchenharvie) and Patrick Warner of the Ardeer Estate. It ran to the port of Saltcoats from Ardeer, and had a number of short branches to different collieries. When it was built, it was said to be the 'most complete water system of colliery transport ever devised in Britain'.

The canal was built by Cunningham and Warner and other coal owners, in part to avoid paying the tolls that were charged on the road leading to Saltcoats harbour, but also because the soft sandy ground made it difficult for horses to haul the heavy coal waggons. The part of the canal near Ardeer was built along the line of bogs and lochs that were the remnants of the River Garnock and used to make Ardeer an island. Auchenharvie was situated at what used to be the mouth of the Garnock. Patrick Warner had previously dug a drainage ditch, known as the 'Master Gott', to reclaim the bogs and lochans on his Ardeer Estate, and sections of this were incorporated in the canal.

Saltcoats harbour had been built by Robert Cunninghamme when he had developed coal pits on his estate and established salt pans, using the coal to produce salt that he exported from the harbour. Cunninghame was greatly respected in Ayrshire for the contribution he made to the economic growth of the county.

The canal was 2.25 miles (3.6 kilometres) long, 13 feet (4 metres) wide at the waterline, 12 feet (3.7 metres) wide at the bottom and 4 feet (1.2 metres) deep, but deeper and wider in places because of the lie of the land, and the sides were angled at 45 degrees. It was fed by the Stevenston Burn, where a spill dam controlled the level, and by water pumped from various collieries.

John Warner, Patrick Warner's brother, supervised the construction. Heavy puddled clay was brought to the workings, with difficulty, by cart to make the canal hold water. The canal took just four months to build and its cost of £4,857 4s was borne by the

Stevenston Coal Company. Once in operation, the cost for transporting coal was 3*d* per ton, while the trans-shipment from barge to cart to seagoing vessels was a further 8*d* per ton.

The canal's three branches were within what is today Ardeer Park; one ran as far as Ardeer House on a circuitous route to avoid a stone quarry. The eastern end of the canal branched, with one spur running to a colliery in Hill Side Field and the other running to the Bogpit colliery.

The canal opened on 19 September 1772. Eight barges were built for the canal, each able to carry between 12 and 15 tons (12.19 and 15.2 tonnes) of coal, which would have taken fifty horses and carts to haul by road. The Saltcoats terminus was a coal yard, and offices, about 600 yards (550 metres) from the harbour, at a site that is still known as Canal Street. The shott at the harbour was very hard igneous rock and the cost and effort of cutting the canal through it was not worthwhile. For this reason, the canal never connected directly with the harbour.

The Earl of Eglinton obtained the right in 1805 to establish a toll gate to the harbour and to levy charges on coal carts, which at first brought him £30 a year. In 1811, however, the earl increased them tenfold, whereupon Robert Cunninghamme decided to build his own waggonway to avoid the iniquitous toll. He did, initially, using wooden rails attached to stone sleepers, the waggonway being built along the rocks of the foreshore. The earl promptly disputed the ownership of the land and threatened legal action. In addition, local householders complained that the waggonway restricted their access to the seashore. By 1812, the track had reached as far as the Saracen's Head inn but, as the earl failed to pursue the legal case, the waggonway was completed and went into active use. The Stevenston Coal Company owned fifty horses which were used both for hauling the waggons and towing the barges.

The waggonway was apparently later relaid using cast-iron fish-bellied rails. Following an agreement with the earl, the line had reached the coal quay by 1827. The canal probably fell largely out use after the construction of the waggonway, which continued in use until 1852 when the export of coal from Saltcoats finished, and the line was lifted for scrap.

The 1856 OS map shows the probable line of the canal with 'Canal Bank' and a 'Canal Cottage' still marked, the latter probably lived in by one of the canal workers. Even today it is remembered in Canal Street, Canal Place and Canal Crescent. However, in 2014, the only remaining parts of the canal itself lay within Ardeer Park at Stevenston.

It was the building of the Glasgow & South Western Railway that contributed to the elimination of the western end of the canal, and of the waggonway. Today the old coal yard and the site of the canal basin remain as open ground, while a pair of gateposts next to the railway may relate to the waggonway.

Ordnance Survey Grid Reference: NS 25989 42064

WALES
Aberdare Canal

While the Aberdare Canal (Welsh: Camlas Aberdâr) in Glamorganshire struggled to pay its way for most of its eighty-eight-year existence, the Glamorganshire Canal, into which it fed at Abercynon, was, for a time, the most financially successful canal in Britain. Although the Aberdare Canal was built 'on the cheap' by the ironmasters of Merthyr, its design was a tribute to the ability of its engineer, Thomas Dadford. It was, strictly speaking, a contour canal, although it had a rise of some 568 feet (173.1 metres) in its 25.5-mile (41-kilometre) length. Sadly, however, Dadford's skill was not appreciated by his paymasters and he was dismissed before the canal was completed.

It opened in 1812 to serve the iron and coal industries. When railways arrived in the area, the competition did not immediately affect its traffic, but the closure of much of the Welsh iron industry in 1875, and subsidence of the canal due to coal mining, made it uneconomic. The Marquess of Bute took it over in 1885 but failed to halt its decline and, in 1900, it was closed, though the company continued to operate a tramroad until 1944. Most of the route of the canal was buried by the construction of the A4059 road in 1923, but a short section at the head of the canal remains in water and is now a nature reserve. The company was wound up in 1955.

The Aberdare Canal Company had been incorporated by an Act of Parliament in 1793, authorising it to build a canal from Aberdare to Abercynon and a railway from Aberdare to Glyn Neath on the Neath Canal. The Act also empowered the company to build tramroads to any mines, quarries or works within 8 miles (13 kilometres) of the canal or railway. The proprietors had powers to raise an initial £22,500, and a further £11,000 if it was needed.

Despite the legal authorisation of the canal, it was not at first thought to be viable, as the Hirwaun ironworks was, at that time, the only likely user for the transport of raw materials and finished goods. The canal company leased some limestone quarries at Penderyn to stimulate economic activity, conveying the stone by tramroad to the Hirwaun ironworks and to its own lime kilns. By 1806, two further ironworks had opened, encouraging the company to pass a resolution at a meeting held in September 1809 to build the canal. Edward Martin from Morriston was employed to re-survey the route, which he did by 9 January 1810, and Thomas Sheasby Junior was engaged as engineer.

The Glamorganshire Canal Company agreed to waive tolls on all stone and lime cargoes which were for the new canal, and the construction of the 6.75 miles (10.8 kilometres) of canal began. Thomas Sheasby resigned as engineer in August 1811 and was replaced by George Overton, who worked for the company on two days each week.

The total fall of the canal required the construction of two locks, one at Cwmbach with a fall of 9.2 feet (2.8 metres), and the other at Dyffryn with a fall of 3.8 feet (1.2 metres).

Along the canal, there was a total of seven overflow weirs to allow surplus water to return to the river. A feeder from the Afon (river) Cynon supplied water to the canal at Canal Head, an aqueduct carried the canal over Nant Pennar, and a stop lock was built at the bottom end, where the canal joined the Glamorganshire Canal just below its Lock 17. Although Richard Blakemore, who owned the Pentyrch ironworks and the Melingriffith tinplate works, wanted any surplus water to be returned to the river, he was ignored, and the water was supplied, as had previously been agreed, to the Glamorganshire Canal. The Aberdare Canal opened for traffic in May 1812, although some outstanding work was completed over the following months.

The four main carriers on the canal were provided with wharf space at Canal Head, Ty Draw, where four 80-foot-long (24-metre) wharfs were built, and the carriers were also allocated wharf space in the pound above the Cardiff Sea Lock by the Glamorganshire Canal Company. It would have taken a boat three or four hours to make the journey from Canal Head to the junction with the Glamorganshire Canal at Abercynon. Although a round trip, from Aberdare to Cardiff and back, could theoretically have been done in thirty hours, many of the boatmen stayed overnight at Nantgarw, so two round trips a week was usual.

Trade in iron began well, but the disastrous depression – caused by the Napoleonic war – that began in 1813 resulted in the bankruptcy of the Hirwaun and Abernant ironworks, and the Llwydcoed ironworks ceasing production. With no other significant sources of income, the Company had no alternative but to close the canal and abandon the tramway to the Neath Canal.

The economy finally began to recover in 1818, when William Crawshay leased the Hirwaun ironworks. After reconstructing and expanding the works, he gradually bought shares in the canal company to ensure that he would be able to export his products. By 1826, he owned 96 per cent of the shares. Between 1823 and 1826, he had the banks of the canal raised to allow the capacity of boats using the waterway to be increased from 20 tons to 25 tons (20.3 to 25.4 tonnes). Crawshay financed the work by selling eleven of his shares, and the rest of the costs were met from revenue, so no dividend was paid in 1826.

Crawshay's ownership of the canal ensured its stability. Then, its fortunes began to improve. In 1837, the first colliery for the extraction of high grade 'steam coal' was sunk at Abernant-y-Groes by Thomas Wayne, who had previously been the clerk to the canal company. Thomas Powell, an entrepreneur, sank another pit at Tyr Founder in 1840, and hit the rich 'four foot seam' two years later. Between 1840 and 1853, fifteen more pits, including Middle Duffryn, Upper Duffryn and Blaengwawr, all close to the line of the canal, were opened. The next thirty years were a period of general prosperity for the canal as it could rely not only on its staple business from the ironworks, but also shared in the business created by the burgeoning coal industry in the Cynon valley, and a number of basins and tramways had to be built for exporting the coal. Powell developed a system of carrying coal in boxes in the boats as a way to reduce breakage and make the handling of the coal easier. It is probable that he borrowed the idea from the 'box boats' that had been used to bring out coal from the Duke of Bridgewater's mines at Worsley. As a true entrepreneur, he negotiated the right to carry coal, with the empty boxes not incurring tolls on the return journey.

With the increasing traffic, more water was needed, and the ponds on Hirwaun Common were extended to make a reservoir, covering 47 acres (19 hectares), from which water was supplied to the canal by way of the Afon Cynon and the feeder at Canal Head. The company also decided in 1845 to build a pumping engine at Tyr Founder to raise water from the Afon Cynon to the canal just above Cwmbach Lock. The Aberdare Company approached the Glamorganshire Canal Company, suggesting that they share the cost of the engine since they too would benefit from the extra water. The Glamorganshire Company offered to pay two-thirds of the cost, providing that they would own the installation, and this was agreed.

The vast increase in coal production in the Welsh valleys soon attracted railways to the area. The very first was the Aberdare Railway in 1846, which branched off the Taff Vale Railway's line from Cardiff to Merthyr Tydfil at Abercynon, and ran up the valley to Aberdare. Many collieries in the valley were offered links to the railway, but the Aberdare Canal Company refused its consent to the railway building bridges over the canal. It was only when a court ruling was obtained in 1851 that the Wyrfa Coal Company was allowed to build the first bridge, after which others followed.

By 1897, the amount of traffic on the canal had dropped to 7,855 tons (7,981 tonnes), and boats were finding passage along the canal extremely difficult as a result of bridges and the towpath sinking because of subsidence. The cost of maintaining the canal continually increased, and finally the decision was taken, in November 1900, to close it on safety grounds. The company continued to operate the tramroad between Penderyn, Hirwaun and Aberdare, and eventually the Penderyn to Hirwaun section was converted to standard gauge and linked to the Great Western Railway (GWR) in 1904. The remainder of the tramroad was sold to the colliery owners in 1944.

The canal itself lay unused and derelict until 1923, when it was bought by the Aberdare and the Mountain Ash urban district councils. When the Aberdare Canal Act was passed in 1924, it had authorised the councils to purchase the canal, and most of its line was buried under the A4059 and B4275 roads. Oddly, it took until 1955 for the Aberdare Canal Company to be wound up.

Canal Head House at the top of the canal, which was originally occupied by the clerk of the canal company, still exists as a private dwelling.

Ordnance Survey Grid Reference: ST 07892 95458

Bowser's Level

Pinged Marsh in Carmarthenshire is a coastal lowland area of comparatively recent origin. It developed at the mouth of the River Gwendraeth Fawr, behind and to the east of the great dune complex of Pembrey, during the post-medieval period. George Bowser built a short canal, known as Bowser's Level, in 1806, from Pinged Village to a tramroad which crossed the marsh to a wharf on the Gwendraeth Fawr.

The Bowser family had moved to Wales from Newington Green, Middlesex, in the late eighteenth century. George Bowser, who had been born in 1778, became heavily involved in coal mining in the area that stretched from the Gwendraeth Fawr to Dyfatty. He leased land in several places for mining, tramroads and canals. He built the level and a tramway, as well as another short canal from his mines, which included Coed Evan Ddu, Syddin and Brynwtha, to his wharf between Muddlescombe and Holway to export his coal.

In 1816 he also leased the closed Cwm Capel colliery from Lord Cawdor and reopened it. He built another tramroad from this colliery to Carreg Edwig. In 1816, together with Thomas Gaunt and two other partners, Bowser leased land from Lord Ashburnham to build and operate Pembrey Harbour. The four partners had already set up the Pembrey Iron & Coal Company, which leased the Gwscwm colliery in about 1818, as well as the adjacent Furnace Iron Works. However, George Bowser later withdrew from the harbour agreement, considering the total cost of £70,000 to £80,000 to be excessive.

Bowser's Level was mostly disused by 1816, when the Kidwelly & Llanelley Canal (K&LC) offered an alternative route to the sea. It closed completely in 1867 when the K&LC closed.

Ordnance Survey Grid Reference: SN 42802 01108

Burry and Loughor Canals

The River Burry has a very wide estuary, between Worms Head and the southern coast of Carmarthenshire. From the bar at the entrance of Burry Harbour, near Holmes Island, to where the River Loughor joins it, the Burry is about 12 miles (19.3 kilometres) long. Over the bar, which changes frequently, there is about 6 feet (1.8 metres) at low water, with between 3 and 5 fathoms (5.5 to 9.1 metres) in the harbour.

According to Joseph Priestley, the River Loughor rises in the mountains south of the town of Llangadoc and flows to Llangennech Ford, which is the head of navigation. The Lliedi, a small stream, joins the River Burry. The only Act of Parliament concerning these rivers is 'An Act for the improvement of the navigation of the Rivers Burry, Loughor, and Lliedi.' It appointed commissioners to enlarge and deepen the rivers, to erect buoys, beacons, and lights, and to be responsible for pilots, anchorages and mooring. Those eligible to be commissioners had to be freeholders of land, owners of mines or works, or shipowners who lived or had investments in industry within 7 miles (11.2 kilometres) of one of the rivers named. The business of the commissioners was to be responsible for appointing a committee of five members annually.

Records show that coal mining began in the Burry valley as early as 1540 but transport, which depended on the Gwendraeth Fawr at the time, was possible but treacherous.

However, in the nineteenth century growing demand for coal, limestone and iron ore meant that navigation in the river needed to be improved and harbourage increased. A few years after the harbour at Pembrey opened, another harbour was built at Burry Port. By 1840, the new harbour, which was fed by a number of canals and waggonways, finally offered an efficient way to export Gwendraeth coal by sea although, at the time, there was no village or town of Burry Port. The earliest records of Burry Port as a town appear around 1850. The importance of the new town was clear when the railways reached Burry Port, with the station serving both Pembrey and Burry Port.

So great was the volume of coal from the whole of the Gwendraeth valley that the canal network was unable to handle it, and part of the network was quickly converted into the Burry Port & Gwendraeth Valley Railway as the port continued to grow in importance and the volume of shipping increased. Traffic on the canalised Burry and Loughor rivers probably finished soon after the opening of the railway and floating dock in 1833.

Ordnance Survey Grid Reference: SN 44458 01522

Cemlyn Canal

Unusually for north-west Wales, where there are few canals, the Vale of Ffestiniog has two of them, one of which is the very short Cemlyn Canal, which connects the River Dwyryd to the quay of the former Diffwys quarry at Maentwrog. The other, from Porthmadog to Tremadoc, is thought to have been navigable, and even to have dispatched a steam-powered ship to New York at some time in the 1830s. There is certainly a building that could have been a warehouse on the site of the canal's basin at Tremadoc (which is now infilled), and the canal passed the ironstone mine at Pen Syflog. On the whole, however, it is more likely that it was intended as much for drainage as for navigation.

As the Dwyryd is followed upstream, immediately above Cemlyn Quay is what appears to be a large ditch; this is the Cemlyn Canal. It was created by realigning a stream which flows through part of Maentwrog. The canal had a quayside some 312 yards (285.2 metres) long, which included Parry's Wharf. It had no gates and was not wide enough for vessels to pass, and depended for its accessibility on the flow of the tide. By 1840 there was a bridge over the canal, which is still there today. The disadvantage of the bridge was that boats rigged with sails had to lower their mast to pass up the canal. The majority of boats on the Dwyryd appear to have carried a single mast, but with a variety of rigging.

Some idea of the quantity of slate – the principal cargo – being produced at that time can be seen from figures for shipments on the Dwyryd. Between 1800 and 1865, 192,288 tons (195,373.6 tonnes) of slate were shipped from the Cemlyn Canal, Cemlyn Quay, and Parry's Wharf. Assuming that the cargo capacity of the average boat was 6 tons (6.09 tonnes), it would mean that around forty-three laden boats would depart from Cemlyn each month. The local boats would have to transfer their cargoes of slate to sea-going vessels in the offing. On some days, the river must have been a very busy place. Between 1800 and 1868 it is known that 420,778 tons (427,530.1 tonnes) was shipped on the Dwyryd. Before 1836, the boatmen taking the slate from Cemlyn out to the ships were charging 15s per ton for doing so.

Once again, this is a story of a canal being superceded by a railway but, in this case, the railway was the narrow gauge Ffestiniog Railway. It was a working railway for carrying slate, not the tourist attraction it became in the twentieth century. It had started by charging 6s per ton when the trains were horse-drawn. But in 1864, a year after the introduction of steam locomotives, the cost fell to 2s 6d per ton.

Ordnance Survey Grid Reference: SH 63974 39706

Clyndu Canal

The Clyndu underground mining canal at Landore, in the lower Swansea valley, was most remarkable; it was connected with underground railways, and was built by Lockwood, Morris & Company. It was constructed in around 1747 beneath Graig Trewyddfa at Morriston, and it has claims to having been the world's longest canal of its type.

Today, Landore (Welsh: Glandŵr) is a mainly residential area of Swansea. It is located about 2.5 miles (4 kilometres) north of the city centre. The first copperworks in the Swansea area was opened in Landore in 1717, and in the 1860s Carl Wilhelm Siemens perfected the open hearth furnace at a local works. By 1873 the area had one of the world's largest steelworks, and the amount of industrial pollution in Landore inspired the doggerel:

> It came to pass in days of yore
> the Devil chanced upon Landore.
> Quoth he, "By all this fume and stink
> I can't be far from home, I think."

The village of Clyndu now only exists in the name of Clydu Street in the Morriston area of Swansea. There is no point-of-access to the former mine, and no means of verifying whether any evidence of the canal remains.

Ordnance Survey Grid Reference: SS 66082 95649

Crymlyn Canal

The Crymlyn bog is the most extensive lowland fen in Wales, and its survival is all the more remarkable because what is now a fascinating nature reserve is surrounded by a dense industrial landscape. The bog lies on the eastern edge of Swansea, where it rests in a large depression gouged out during the last Ice Age.

The Crymlyn Canal in Glamorganshire has also been referred to as 'Clawdd-y-Saeson', or the 'Englishman's Ditch'. It was probably first dug in the Middle Ages as a drainage ditch. Later excavations in the bog were reported to have found the skeleton of a boat, but no remains or detailed records exist to establish the find.

The canal is now no more than a drain.

Ordnance Survey Grid Reference: SS 69415 94754

Cyfarthfa Canal

Anthony Bacon, a native of Cumberland, and his business partner William Brownrigg built iron furnaces at Cyfarthfa in 1765, and it was probably they who built a small tub boat canal in 1770, which was 2 miles (3.2 kilometres) long on the level, between the nearby coal pits and the ironworks. By 1806, the Cyfarthfa ironworks had become the largest in the world because it had been the first in the area to change to the production of bar iron and other advanced processes in the late eighteenth century. In the second half of the nineteenth century, the works switched to the production of steel.

The Taff valley to Merthyr Tydfil was an important transport route that was followed by the Glamorganshire Canal and the Cyfartha Canal, as well as roads, tramroads and railways. The valley's earliest form of communication had been the road. Today, many features of the former industrial railways survive, including embankments, cuttings and bridges, together with those of the Glamorganshire Canal, although the Cyfarthfa Canal is now untraceable. The present appearance of the landscape is due largely to the rapid growth of the iron and coal industries in the eighteenth and nineteenth centuries, including coal and ironstone extraction and the iron and steel industry.

The Cyfarthfa Canal, which was shown on Yates's map of 1799, was connected to an adit at Wern, from which coal was loaded directly into small iron boats, locally called 'buckets'. Anthony Bacon supplied iron cannon to the Board of Ordnance, and from 1777 the entrepreneur Richard Crawshay was a partner in the business. The casting of cannon was done at the Cyfarthfa ironworks. Bacon was elected a Member of Parliament in 1782 and had to end his involvement with government contracts because of the clash of interests. Richard Crawshay then leased the Cyfarthfa works, after which the ironworks and the coal and ironstone workings expanded rapidly. The canal appears to have been abandoned in about 1836. The works remained in the Crawshay family after Richard's death on 27 June 1810. They then passed to his son William Crawshay I, who directed operations from his office in London, leaving the day-to-day management, including the great expansion of the works, to Richard's grandson William Crawshay II.

Ordnance Survey Grid Reference: SO 03481 06761

An exposed part of the Cyfarthfa Canal. Image used by permission of Ian Lane.

Doctor's Canal

It was Dr Richard Griffiths, a Welsh industrial pioneer, who introduced Rhondda coal, from a pit at Hafod near Newbridge (today's Pontypridd), to the world in the late eighteenth century, and who thought of linking it by canal and tramroad to the Glamorganshire Canal.

Griffiths was a member of the Gellifendigaid family of Llanwynno; he was the third child, and second son, of William and Elizabeth Griffiths. Richard practised medicine in Cardiff, but it was through his family connections that he began prospecting for minerals and for which he became best known. He was described as a 'dynamic and colourful personality' and a practical joker, even, on one occasion, arranging a mock funeral for himself.

His youngest sister was married to Evan Morgan, who owned the Hafod Fawr Estate, a farm in the Lower Rhondda near modern day Trehafod. In 1808 Griffiths obtained a lease on the mineral rights for the farm from his brother-in-law. In 1809, he made two sub-leases of the rights and allowed Jeremiah Homfray to open a level below the estate, on the east side of the River Rhondda. Homfray continued to work the level until he was declared bankrupt in 1813.

To make the Hafod Fawr Estate more profitable, Griffiths decided to improve transport links to the newly-opened Glamorganshire Canal, which linked ironworks at Merthyr Tydfil to the port of Cardiff, both of which represented eager markets for coal. The existing means of transporting coal was using pack horses, so Griffiths first built a tramroad from the Hafod Estate to Newbridge, where he bridged the River Taff. The tramroad, which used horse-drawn waggons, opened on 29 September 1809.

Some years later, Griffiths suggested building a branch canal and tramroad from the Glamorganshire Canal at Denia to a colliery at Treforest. The canal, which he built at his own expense and which became known as the Doctor's Canal, was 1 mile (1.6 kilometres) long on the level. It was opened in 1813 and was joined to the Glamorganshire in 1813 after some delay, caused mainly by the Glamorganshire company's concerns about the Doctor's Canal's water supply.

The link between Griffiths' branch and the Glamorganshire Canal proved most valuable when Walter Coffin, who was the first to sink a deep mine in the Rhondda, was granted rights to use Griffiths' tramroad. As Coffin's pit was further up the valley at Dinas, he had to build his own 4-mile-long (6.4-kilometre) tramroad to connect his colliery to Griffiths' branch canal at Trehafod.

On Griffiths' death, the rights to the Trehafod Estate passed to his family and the rights to deep-mine Trehafod colliery were let to John Calvert in 1851. Griffiths' tramroad continued to be used until the arrival of the Taff Vale Railway's branch to the Rhondda in 1841. Traffic on the Doctor's Canal declined, although some of the tramroad and the canal remained in use. Ownership passed in 1826 to a nephew of Dr Griffiths, Revd George Thomas, and his brother. It was disused by 1914 and derelict by 1918.

Ordnance Survey Grid Reference: ST 04854 90948

Earl of Ashburnham's Canal

In the mid-seventeenth century, the Ashburnham family were modest Sussex gentry but, within less than a hundred years, they had acquired an earldom and extensive estates in Bedfordshire, Suffolk, Lancashire, Wales and elsewhere. Their success was due to them making good marriages, and the fact that they were shrewd businessmen. The estates in Breconshire and south-east Carmarthenshire came through the marriage of John, who was created first Lord Ashburnham in 1689, to the only daughter of Walter Vaughan of Porthamel House, near Talgarth.

On the Pen-bre Estate of the Vaughans, Lord Ashburnham ordered trial workings to discover coal. When large deposits of coal of good quality were found, pits were sunk at Coed y Marchog and Coed Rhial on the western slope of Pen-bre mountain. The coal was carried by pack horse to the estuary of the River Gwendraeth and to the Burry inlet, and from there it was shipped mainly to the west of England and to Ireland.

It was John, the second earl, who, impressed by the success of Kymer's Canal, built by Thomas Kymer from 1766 to 1768 to carry coal from the Gwendraeth Fawr to Kidwelly, put forward a scheme for building a canal from the foot of Pen-bre mountain to the Gwendraeth estuary. The scheme failed to get off the ground because of the opposition of the miners at his pits. The earl wryly observed, in a letter to his agent in 1770, that their opposition was based largely on self-interest, since a canal would mean that the earl would not pay for their horses to carry the coal.

It was not until the 1790s that the earl was able to proceed with his scheme. The first length of the canal, starting below Coed farm close to the Llandyry–Pinged road, probably began in 1796. The colliery account book records a payment to Griffith Jenkins & Company that year for cutting the canal, the purchase of barges, the construction of a tramroad to the canal, and a towpath. By 1799, the canal had crossed the Kidwelly–Pen-bre road near Saltrock farm and had reached, and been cut through, a bank of earth at the edge of the marsh.

In the summer of 1801, different contractors were constructing it through the Pinged marsh. By the end of that year, the canal had reached its terminus at Pill Towyn, a creek that ran in from the south bank of the Gwendraeth Fawr. On a short fork at the head of the creek, two shipping places were built, one of them called 'Black Creek'(Welsh: Pill Ddu).

Pinged Marsh had developed at the mouth of the Gwendraeth Fawr, behind the sand dunes at Pembrey, during the post-medieval period. The dunes protected the area from the sea but impeded the drainage of the land behind them. In the early seventeenth century, a survey mentioned 'the marsh on both sides of the bridge called Pont y Spowder', (Spudder's Bridge), the late medieval stone bridge that still survives. Once reclaimed, the land became common pasture. The north side of the marsh had been drained by

1766, when the industrialist Thomas Kymer built his canal from Pont-iets to Kidwelly. Nevertheless, much of the area was still marshland and subject to regular inundation into the early nineteenth century. Much of Kidwelly's maritime trade used Frankland Quay, which used to be on the Gwendraeth Fawr, 875 yards (800 metres) south-west of Spudder's Bridge.

In 1806, George Bowser built his short canal from Pinged Village to a tramroad which crossed the marsh to a wharf on the Gwendraeth Fawr. A third canal was excavated by the contractors Pinkerton and Allen in 1814–24, for the Kidwelly & Llanelly Company, from a junction with the Earl of Ashburnham's Canal, through Pinged Marsh, to Frankland Quay. It was also joined by Bowser's Canal and had a branch westward over an aqueduct to Trimsaran (which was replaced by a railway line in 1865).

The Earl of Ashburnham's Canal was about 2 miles (3.2 kilometres) long with no locks, although a number of accommodation bridges were built across it. In 1805, a short branch towards Ffrwd was added when new coal levels were opened in Coed Rhial. The entrance to Pill Ddu was deepened in 1816 and a dry dock was added in 1817.

William Raynor, a shipwright from Carmarthen who settled in Kidwelly, began by building and repairing barges for Kymer's and the Earl of Ashburnham's canals. He leased a length of the river bank in 1803 for a shipyard, from which he launched brigs and sloops, the largest of which was the square-sterned brig *Margaret* of 163 tons (165.6 tonnes) burden built in 1814.

In the days before the Welsh religious revival, Kidwelly had a number of inns and pubs, including: The Pilot Boat, The Jolly Sailor, The Hope and Anchor, the Ship and Castle and the Yarmouth Arms. Today it is difficult to believe that it was once a port of such importance.

Signs of the decline of the area were beginning to appear in the first decade of the nineteenth century. The movement of sand was causing navigational difficulties. At the end of the eighteenth century, the channel of the Gwendraeth River at the mouth of the Towy was deep enough for ships drawing 16 feet to 18 feet (4.8 metres to 5.4 metres) of water to get into Kymer's dock, but by 1801 it was becoming shallower because of the build-up of sand. A new channel solved the problem for a while but, by 1809, it too was becoming shallow, in spite of the rise of the spring tides.

The Kidwelly & Llanelly Canal & Tramroad Company, formed in 1812, tried to restore the channel and to improve and extend the canals serving the Gwendraeth valley. The directors of the Kidwelly Company called in John Rennie in 1819, but his report the following year was not encouraging. He did suggest extending the canal to serve one or more collieries, while pointing out that its cost would be too much for what would be no more than a temporary improvement.

By 1836, the sands had shifted again, making it easier for ships to enter Kymer's dock. Four Kidwelly ship's masters said that they believed that vessels of up to 400 tons (406.4 tonnes) burden, with a draught of 16 feet (4.8 metres), would be able to enter the dock on a four-hour flood tide. However, two years later the canals of the valley were linked with the new Burry harbour, where there was deeper water, and the bulk of the coal traffic left Kidwelly for Burry Port.

The Earl of Ashburnham's Canal continued to be maintained and repaired, and shipments of coal were made from it up to 1818. However, on 10 January 1818, John Hay, who was responsible for the working of the collieries at Coed y Marchog and Coed Rhial, and the operation of the canal, wrote to the earl to tell him that the collieries were almost exhausted, and shortly afterwards the canal seems to have gone out of use.

Ordnance Survey Grid Reference: SN 42732 03827

General Warde's (Dafen and Yspitty) Canals

Chauncey Townsend, the son of Jonathan Townsend, a brewer in London in the early eighteenth century, held government contracts for provisioning troops and settlers in Nova Scotia which made him a considerable fortune. Townsend built a canal in Glamorganshire in about 1769 to transport coal from pits that he owned to the Lougher estuary, from where it could be exported. Essentially, this canal was a canalised creek, 0.75 miles (1.2 kilometres) long, from Tireinon and Llwynhendy to Dafen Pill.

In 1801, the Dafen Canal and associated canals built by Townsend were bought by Major-General Warde, who was an entrepreneur. This and the other canals were later named after Warde. The Dafen Canal was probably disused by 1808.

The Yspitty Canal, through the lands of Heol Fach and Pwll Cefn in Carmarthenshire, was built by John Smith for carrying coal, culm, timber, stone and other goods. In January 1786, Smith obtained a deed that granted him these lands, and gave him permission to build a 'navigable cut or canal through the common called Dole Vawr y Bynie, in the hamlet of Berwick … for the carrying of coals from part of the said lands to be shipped at the River Burry'. The coal and other goods were transported by small barges on the canal to wharfs at the mouth of the Loughor river at Yspitty Bank, where they were stored before being loaded on to larger sea-going vessels.

John Smith was also authorised by the deed to enlarge his existing canal and extend it westwards. He proposed to open a new colliery near Dafen Bridge, and intended to improve the facilities for shipping the coal and other goods at Yspitty.

In January 1797, Smith's sons, Charles and Henry Smith, jointly inherited the Yspitty Canal. In January 1801, the brothers sold their interests in the canal to Major-General Warde.

The Yspitty Canal was disused by about 1829. A group of local businessmen, who had interests in the Llangennech Coal Company, formed the Llanelly Dock Company at about the same time. The company planned to develop the Llangennech coal seams in the Dafen district of Llanelli, and to build a 2-mile-long (3.2-kilometre) railway line to connect with the new dock on Llanelli Flats. An Act of Parliament authorising the construction of the line and dock received the royal assent on 19 June 1828. Various problems delayed the opening of the railway, which in part used the route of the Yspitty Canal, until 1833. The line, which was known simply as the Dafen branch, survived for well over a century and closed in 1963. Some of the route of the railway can still be traced, although there are no remains of the canal.

Ordnance Survey Grid Reference: SN 52865 01348

Giant's Grave and Britton Ferry (Jersey) Canal

Access to Neath for coastal vessels of up to 200 tons (203.3 tonnes) burden had been made easier by the construction of the Neath Navigable Cut in 1791. A second Neath Canal Act was passed on 26 May 1798, authorising an extension of about 2.5 miles (4.0 kilometres) to Giant's Grave, where better means of loading goods aboard seagoing vessels were available. Thomas Dadford surveyed the route, but Edward Price, from Govilon near Abergavenny, acted as engineer. This section of the canal was paid for by Lord Vernon.

The extension was completed on 29 July 1799, and terminated at a basin close to Giant's Grave Pill. Flood gates on the canal enabled water to be released into the pill to scour it free of silt. The total cost of the project was about £40,000, which included nineteen locks and a number of tramways.

Between 1815 and 1842, more docks and wharves were built at Giant's Grave, and the canal was extended to Briton Ferry by the construction, in 1832, of the Jersey Canal, which was just over half a mile (1 kilometre) long, and was built by the Earl of Jersey without an Act of Parliament. A further short extension was made around 1842. The final length of the canal overall was 13.5 miles (21.7 kilometres).

Ordnance Survey Grid Reference: SS735945

Glamorganshire Canal

'Navigation House' at Abercynon pre-dates the village. It was originally built by the Glamorganshire Canal Company at the terminus of a tramway where, in 1804, Richard Trevithick demonstrated the world's first steam locomotive running on rails and hauling a load of 10 tons (10.1 tonnes).

The Glamorganshire Canal, 25 miles (40 kilometres) long, ran from Merthyr Tydfil to Cardiff. Its construction began in 1790 and the canal was fully open by 1794. Its main purpose was to transport the raw materials and finished products of the Merthyr iron industries. At Nantgarw there was also a china and porcelain works, which would not have been economic, because of breakages, if it had depended on the roads for transport. The smoke and sweat and the sheer industry of these works must have made a scene that is hard to imagine today. The canal could carry barges of 20–25 tons (20.3–25.4 tonnes) burden. Ultimately, it would be a great financial success, but there were difficulties, not least that a journey along the full length of it was slow because of its numerous locks. In the Abercynon section, for instance, there is a rise of 200 feet (60.9 metres). The locks that still exist are impressive to say the least.

The combination of iron ore, coal and limestone to be found in the area around Merthyr Tydfil resulted in a number of industrialists being attracted to the area in the second half of the eighteenth century, one of whom was Richard Crawshay. Between 1759 and 1784, four major ironworks, at Dowlais, Plymouth, Cyfarthfa and Penydarren, had begun production, but transporting the finished products was difficult on the steep and poorly made roads down the valleys.

In 1786, after Crawshay and his partners had acquired the lease on the Cyfarthfa ironworks, they engaged Thomas Dadford, the canal engineer, to survey a route for a canal to Cardiff. The survey, paid for by Crawshay and three other ironmasters, was completed in 1789. With the support of Lord Cardiff, who was the most powerful landowner in the region, a bill to authorise the Glamorganshire Canal and constitute a company of proprietors passed through Parliament without amendment, and the Act was granted on 9 June 1790.

The Act's well-tried formula authorised the proprietors to raise £60,000 in capital by the issue of shares, and a further £30,000 by mortgage if required, to build the canal and any such branches or feeder tramroads as required, to link it to any works within 4 miles (6.4 kilometres) of its course. The tramroads were regarded as part of the canal itself, and the company was therefore able to acquire land for them by compulsory purchase if necessary.

Construction began from the Merthyr Tydfil end in August 1790, supervised by Thomas Dadford, his son Thomas Dadford Jr, and a team of workmen. The Merthyr to Newbridge (later renamed Pontypridd) section was completed in just under two years, being completed in June 1792. The rest of the canal was opened in stages, first to Pwllywhyad in January 1793 and then to Taffs Well by June that year. By this time, the canal had greatly exceeded its budget. Corners had to be cut with the result that, although the final section to Cardiff was opened on 10 February 1794, it was poorly built, so much so that there were a number of stoppages during the rest of the year while repairs were made.

Water for the top of the canal came from the tail-races from the Cyfarthfa ironworks. This water had previously been fed back into the River Taff, so that it could be reused by the Plymouth ironworks. In order to safeguard this supply, all the water discharged from the canal's third lock was meant to be directed into the Plymouth feeder, rather than into the canal below it. Legal action was instituted by both sides for some years, with the true winners being the lawyers.

To make matters worse, there was a breach in December 1794, but Dadford refused to even begin making repairs without payment in advance. When no money was forthcoming, he dismissed the workers. The company tried to recover £17,000 from the Dadfords, father and son, and went so far as to have them arrested. The engineer Robert Whitworth was called on to adjudicate, and he employed two independent surveyors whose findings, by and large, were in favour of the Dadfords, and they refunded only £1,512 to the company.

A planned branch to the Dowlais and Penydarren ironworks, which would have had a rise of 411 feet (125 metres) in only 1.75 miles (2.8 kilometres), was dropped in favour of two tramroads, one from each works. The canal's main line had a rise of about 542 feet (165 metres) and required fifty locks: an average of two per mile. It followed the western side of the valley down to Abercynon, where it crossed the River Taff on an aqueduct and then followed the eastern side for most of its route to Cardiff.

A second Act of Parliament was passed on 26 April 1796, authorising the extension of the canal by 0.5 mile (0.8 kilometre), meaning it ended at a sea lock in the docks at Cardiff. The extension was opened in June 1798 with a naval review and the firing of ships' guns. The total cost of the canal had been £103,600.

Richard Crawshay, the largest shareholder, tended to treat the canal as his own personal property. The other ironmasters retaliated by proposing a tramroad from Merthyr to Cardiff in competition. Crawshay agreed reluctantly to somewhat reduce the tolls on the canal, but the ironmasters whose works were on the east side of the valley built their own tramroad to Merthyr, which opened in 1802. The Plymouth ironworks also benefited from the opening of the tramroad, as traffic on the upper section of the canal was reduced, meaning that less water was used by the locks.

Dividends were limited to 8 per cent by the original Act of Parliament, so between 1804 and 1828 the profits were used to give refunds to the traders who used the canal. Alternatively, the company would charge no tolls for a period, and at other times would reduce tolls to a quarter of the rate fixed by the Act. By the 1830s, there were 200 barges at work on the canal.

It must not be forgotten that smuggling was still rife in the Bristol Channel, and the canal was a useful means of distributing the contraband. In 1818, for example, a large shipment of spirits was landed at Barry Island – then still a separate island and a known haunt of smugglers. The preventative men succeeded in seizing three kegs (about 45 gallons), but most of it found its way to the back streets of Cardiff and on to the canal barges, no doubt to soothe the palate of many coal miners and foundry workers. A small number of canal constables were appointed to keep order and help to prevent such pilfering. In 1857, Superintendent Stockdale, who was in command of the Cardiff Borough Police, reported that 'the Glamorgan Canal Company pay and employ two constables under my command, for duty on the canal. They also have a separate police station at the Old Sea Lock [in Cardiff docks]'.

Railways, as opposed to tramroads, arrived in 1841, when the Taff Vale Railway opened to Merthyr. As long as the ironworks on its banks continued in business, the canal maintained its profitability. However, when the ironworks started to close in the 1870s, profits tumbled. In 1876, the canal company was unable to pay the full 8 per cent dividend for the first time.

In 1885, the canal was sold to John Patrick Crichton-Stuart, the Marquess of Bute and the wealthiest landowner in Wales, who made some improvements at the Cardiff end of the canal. However, by then there was not one, but six railway companies serving Merthyr, all competing for traffic. The upper sections of the canal, particularly the 4-mile-long (6.4-kilometres) pound at Aberfan, was suffering from severe subsidence resulting from coal mining. After an inspection, it was decided that the canal must close from Merthyr to Abercynon to safeguard the village, which it did on 6 December 1898.

Traffic on the rest of the canal continued to decline and, when a breach occurred at Cilfynydd in 1915, the company decided against investing in repairs. Instead, a wooden flume was built past the breach, so that the rest of the canal could still receive water from the Elen Deg feeder. When a further breach occurred near Nantgarw on 25 May 1942, engineers examined it but decided not to carry out any remedial work.

Cardiff Corporation agreed to buy the canal for £44,000. This was incorporated in the Cardiff Corporation Act passed in August 1943, allowing the corporation to assume control of the canal on 1 January 1944. On that day, the canal was immediately declared closed, after which it was progressively infilled. In any case, most business had effectively finished in 1942. Section 27 of the Act prevented the closure of the final mile above the sea lock at Cardiff for as long as it was being used by sand dredgers. Two sand and gravel firms continued to use the sea lock and the bottom pound of the canal to carry on with their business.

The end came dramatically on the night of 5 December 1951, when a suction dredger, the *Catherine Ethel*, which weighed 154 tons (156.4 tonnes), crashed into the inner lock gates. The gates collapsed, and all of the water in the 1-mile-long (1.6-kilometre) section drained into the estuary.

Today, some traces of the course of the canal remain, although about half of it is covered by the A470 Cardiff–Merthyr Tydfil trunk road. The section from Tongwynlais to the Melingriffith tinplate works at Whitchurch is still in water and is now a nature reserve. A few bridges and locks still remain towards the head of the canal. There are also some short stretches in water at Nightingales Bush and at Locks 31 and 32 in Pontypridd.

The Canal & River Trust, which took over responsibility for most of the navigable waterways in Great Britain in 2012, says poetically,

> … the people of Wales have a new feeling for their heritage ('treftadaeth' in Welsh). Many remnants of the great Glamorganshire Canal remain, some have been restored, many more wait to be discovered. These remnants and the stories they tell are a real treasure of Britain's historic waterways.

Ordnance Survey Grid Reference: ST 13377 82105

Glan-y-wern and Red Jacket Canals

The Red Jacket Canal was a 3.5-mile-long (5.6-kilometre) canal with no locks, built by Squire Edward Elton of Kilvey Mont, that connected with the Glan-y-wern Canal, and so connected the River Neath with the River Tawe. 2 miles (3.2 kilometres) of it was through the morass of Crumlyn Bog. It was started in late January 1817 with William Kirkhouse as engineer – without an Act of Parliament – and completed in the autumn of 1818.

Richard Jenkins, who owned one half of the Glan-y-wern colliery, was shipping coal from his colliery by horse-drawn carts to Foxhole on the River Tawe. The Turnpike Trust erected a gate near 'The Gwindy' pub, and Jenkins found that with twenty carts a day, each charged a shilling for passing through the turnpike gate, he was paying £1 a day to get his coal to the river wharfs for export. In 1788, he approached Lord Vernon, through his friend Lewis Thomas (Lord Vernon's agent), to ask for a lease to build a canal from his colliery to Trowman's Hole on the River Neath.

Because of the expense of developing the canal, Jenkins took on Squire Elton as an equal partner. Even before Lord Vernon had granted the lease of the land for the canal, Jenkins wrote to Thomas Telford in Gloucester and asked him to come down to begin the canal. On 14 August 1788, Lord Vernon granted a lease to Richard Jenkins to enable him to build the canal, but on the day that he signed the lease Jenkins died. Edward Elton was left with the responsibility of building the canal, which he completed in 1790.

The canal allowed copper ore to be transported directly from the River Neath at Briton Ferry to the Briton Ferry copperworks and the Red Jacket copperworks on the western side of the river. Coal for export was trans-shipped from the small canal boats used on the canal to sea-going ships that were moored at Trowman's Hole.

The name 'Red Jacket' is said to have been applied to the canal because the men who either built or operated the canal wore red jackets, perhaps old military uniforms. The name 'Red Jacket Pill' was later applied to the pill leading to Trowman's Hole.

Richard Jenkins' widow was unable to support her family and signed over the canal and the colliery to Squire Edward Elton for £100. Elton himself died in 1810, still owing large arrears of rent to Lord Vernon, who then served a Notice of Distress on the barges in the canal. Following that, the Red Jacket Canal gradually fell into dereliction until George Tennant took an interest in it in 1817, at the time he acquired the lease of the Glan-y-wern Canal.

Tennant employed William Kirkhouse to extend the Glan-y-wern Canal into the River Tawe at Swansea, making a link between the Neath and Tawe rivers. The canal extension, which was completed in 1818, ran from the Glan-y-wern wharfs, at Red Jacket Pill on the River Neath, to the recently constructed east pier on the River Tawe. Part of the original canal to the Glan-y-wern colliery remained as a branch to the colliery. The Glan-y-wern Canal was disused by 1922.

Ordnance Survey Grid Reference: SS 66338 94040

Gwaun-Cae-Gurwen Canals

The village of Gwaun-Cae-Gurwen, at the head of the Amman valley in the north-western corner of West Glamorgan, at the head of the Clydach River, is 14 miles (22.5 kilometres) from Swansea and 19 miles (30.5 kilometres) from Llanelli. It was the existence of high quality anthracite coal which led to the development of the village in an area of inhospitable moorland and hill pasture.

Because of the isolated position of the area and the distance to the nearest port, although there is some mention of coal-working at Gwaun-Cae-Gurwen in the eighteenth and early nineteenth centuries, it never developed to any great extent.

However, the existence of the Gwaun-Cae-Gurwen Canal indicates that there was a serious attempt to establish a mine, probably around the year 1757. The canal, the oldest in the Swansea valley, was underground – actually a series of canalised levels linked to the surface – which carried both men and coal. Little is known about the extent or the method of working of the canal, either underground or on the surface. The mine at Gwaun-Cae-Gurwen was near the head of the Upper Clydach River, which runs from a poorly drained area south of Cwmgors village, and flows south and south-east through Cwm Gors for about 4.3 miles (7 kilometres) to join the River Tawe at Pontardawe.

It was not until 1837 that Roger Hopkins, a local man, became the first industrialist to realise the potential of Gwaun-Cae-Gurwen. By the time he turned his attention to Gwaun-Cae-Gurwen, he was an experienced and successful engineer. He had founded the Victoria Iron & Coal Company in 1836, and the Swansea & Gwaun-Cae-Gurwen Anthracite Company, to lease and work the coal at Gwaun-Cae-Gurwen. Both companies had registered offices in Bath and the same man, J. J. Skinner, was secretary to both companies.

It was also in 1837 that the Victoria Iron & Coal Company obtained a ninety-nine-year mineral lease at Gwaun-Cae-Gurwen at a rent of 7*d* per ton, subject to a minimum of £583 6*s* 8*d* per annum. This was the equivalent of an annual output of 20,000 tons (20,320 tonnes). It is most likely that that the underground navigable levels were located at Hopkins' Gwaun-Cae-Gurwen mine. Altogether, Hopkins leased 717 acres (260 hectares) of coal and iron ore from the major landowner, Capel Hanbury Leigh of Pontypool, whose family had been ironmasters in Staffordshire before moving to Monmouthshire in about 1570, starting similar works at Monkswood and Pontymoel.

Ordnance Survey Grid Reference: SN 70382 11739

Kidwelly & Llanelly and Kymer's Canal

These canals had a complicated history. The Kidwelly & Llanelly Canal was a canal and tramroad system in Carmarthenshire that was built to carry anthracite to the coast for trans-shipment to coastal ships. Its predecessor, which it later absorbed, was Kymer's Canal, built in 1766, which linked pits at Pwll y Llygod to a dock near Kidwelly. Thomas Kymer had begun mining at Pwll y Llygod and Great Forest (near Carway) in 1760, and he sought parliamentary approval to construct a canal at his own expense in early 1766. The Act was granted on 19 February 1766.

The canal was to run from Kymer's coal pits at Pwll y Llygod on the banks of the Gwendraeth Fawr river to Ythyn Frenig, about half a mile (0.8 kilometre) to the west of Kidwelly, where he built a dock on the southern bank of the Gwendraeth Fach river. The canal was finished and ready to carry traffic in May 1768. Kymer's Canal, which was 3 miles (4.8 kilometres) long, was the first canal of any length in Wales, and was built to carry coal from the Gwendraeth Fawr to Kidwelly, where a quay was constructed. It was later incorporated into the Kidwelly & Llanelly Canal, and subsequently a branch of the Burry Port & Gwendraeth Valley Railway was laid on the towpath in 1873. A length of the Kidwelly & Llanelly Canal can still be seen at Muddlescombe, and at Pwll y Llygod a fine tramroad bridge crosses the canal.

The canal company had a series of meetings in 1811 to consider how to develop the canal. The engineers, Edward Martin and David Davies, who were working together on the Pen-Clawdd Canal, proposed to extend the canal to the top of the Gwendraeth valley, together with another extension which would cross Pinged Marsh to terminate at Llanelli. Their proposal would take the canal through Pembrey, thus providing access to the harbour there.

Davies and Martin's proposal was adopted by the committee, and an Act of Parliament was passed on 20 June 1812 creating the Kidwelly & Llanelly Canal & Tramroad Company. The Act envisaged an upper terminus beyond Cwmmawr at Cwm y Glo, together with a series of feeder canals or tramroads to connect the pits and levels, where coal was being mined, with the canal. It listed thirteen collieries which would be served by the canal. Wharfs at the Llanelly terminus would be built next to the dock, which was owned by the Carmarthenshire Railroad Company (CRC). A lock and weir would be constructed at the point where the canal would cross the existing Ashburnham Canal.

Access to the Burry Port dock, however, became more difficult as the estuary gradually silted up, so an extension to Llanelli was authorised in 1812. The construction of the canal was slow, and it was not until the 1820s that it was linked to a harbour at Pembury built by Thomas Gaunt, a temporary solution until the company's own harbour at Burry Port was completed (which it eventually was in 1832). Tramways owned by the company served a number of collieries to the east of Burry Port.

The company was required by its 1812 Act to finish the extension from Pwll y Llygod to Pontyates, and the first section of the Llanelly branch to the point where it crossed the Ashburnham Canal, within six years. The engineer was Anthony Bower and the contractor was Pinkerton & Allen. The canal reached the new aqueduct across the Gwendraeth Fawr in 1815, but little progress was made between then and 1817. Much of the capital raised had been spent on attempts to reopen Burry Port harbour but with little success.

A second Act of Parliament was obtained on 28 May 1818, which extended the time limit for completing the canal and removed the requirement to clear the river channels to the harbour. In 1819, John Rennie made a report that was not encouraging. He did put forward a scheme, but he pointed out that its cost would be too heavy for what was likely to be only a temporary improvement. Pinkerton & Allen continued with construction, and completed the line to Pontyates and the Ashburnham Canal crossing in 1824. Meanwhile, the company's directors consulted the engineers John Rennie and Edward Bankes regarding the matter of a suitable harbour. Rennie suggested extending the canal towards what became Burry Port, and the construction of a new harbour on the sands at Tywyn Bach.

In 1832, the West Country engineer James Green was asked to advise the company on extending the canal, and he suggested building a line with three inclined planes to reach Cwmmawr, further up the Gwendraeth valley. Green had a great deal of experience with building inclined planes on canals, particularly in Somerset. Surprisingly, on this occasion he underestimated the cost and was dismissed in 1836. However, the canal company seems to have raised more capital from its shareholders, because the canal was finished the following year.

Once it was open, the canal was reasonably successful, and the shareholders received dividends from 1858. In 1865, the company changed its name to The Kidwelly & Burry Port Railway (K&BPR). The following year, the K&BPR amalgamated with the company that ran Burry Port.

Ordnance Survey Grid Reference: SN 46626 09205

Kilgetty Canal

The Kilgetty Canal, in Pembrokeshire, was built in about 1790 by miners from Lord Milford's colliery at Kilgetty near Saundersfoot. It was about 1 mile (1.6 kilometres) long and probably intended to be used by tub boats from the colliery to the coast. It was never finished and probably never used.

While some of the coal produced was sold locally, Lord Milford wanted to reach a wider market. Coal from the mines at Kilgetty, and from further afield, was exported by coastal vessels from Wiseman's Bridge, but as this was only possible in late spring and summer, it meant that the miners were laid off in the winter.

Another attempt at increasing the coal trade was a canal from the Kilgetty Great Pit to Wiseman's Bridge, but no construction seems to have been carried out. Later, a Mrs Child suggested building a tramroad from the Begelly Collieries to Saundersfoot. This would have been easier than building a canal but, again, the idea went no further.

Eventually, in 1829, the Saundersfoot Railway and Harbour Act was passed by Parliament, which also authorised a line from Saundersfoot to Thomas Chapel. The 4 foot (1.2 metres) gauge line was officially opened in March 1834. The trucks, of the type used in the collieries, were horse-drawn until 1874, but in that year a 0-4-0 saddle-tank locomotive was acquired. She was later named *Rosalind* after the daughter of Charles Vickerman, who was the main mine-owner in the local coal industry. The Tenby & Begelly Coal Company had two small vessels, the *Peggy* and the *Mary Anne*, and by 1836 there was a Begelly Wharf at Saundersfoot.

Ordnance Survey Grid Reference: SN 12660 07285

Llansamlet (Smith's) Canal

The Llansamlet Canal was a 3-mile (4.8-kilometre) canal completed in around 1784, stretching from collieries at Llansamlet, Glamorganshire, to the River Tawe at Foxhole, near Swansea.

The present-day village of Llansamlet grew up around the old village cross only after about 1790, when a turnpike road was constructed between Neath and Morriston, with a new bridge across the River Tawe at Wychtree. Llwyncrwn Road, in the village, follows the route of Townsend's waggonway, which was built in 1750 by Chauncey Townsend, the coal-owner. Horse-drawn waggons were used to carry the coal down from his pits to the Upper Bank leadworks and the wharfs at Foxhole for loading on to ships. Later in the eighteenth century, Smith's Canal replaced the waggonway, which can still be traced as a rough track for much of its length.

There were three copperworks in Llansamlet Lower: White Rock, Middle Bank and Upper Bank. After they were built, there was greater demand for coal for the smelting. The coal came down from pits in Llansamlet Upper using Smith's Canal. The old Cnapgoch mansion and its outbuildings near the White Rock copperworks were converted in time to form twenty-four dwellings for workers at the copperworks, while the agent or manager was given a newly-built house near the works.

The Llansamlet Canal was built by John Smith, who was Chauncey Townsend's son-in-law and the owner of collieries at Llansamlet. The canal, which took his coal 3 miles (4.8 kilometres) to Foxhole, began near Gwernllwynchwith House and served a number of other collieries along its route. According to a contemporary source,

> the traffic over it was directed to the wharf at Birchgrove, the spelter works of the Dillwyns [spelter is a zinc metal, sometimes mixed with lead, used to create statuary and decorative tableware], the Middle Bank works of the Grenfells and White Rock works as well as some smaller works.

John Smith's sons, Charles and Henry, inherited the collieries and the canal. In 1816 they built a tramroad, known as Scott's Railway, parallel to the canal from Scott's Pit, above Llansamlet, to Foxhole. The tramroad was later incorporated into the Swansea Vale Railway, which opened in 1852. It is most likely that use of the canal declined and it eventually became disused as a result of the construction of the tramroad. Part of the canal alongside the river at Foxhole was bought by the Swansea Vale Railway in 1855 for use as their railway wharfs.

Today there are a few traces of Smith's Canal to be seen, principally the basin of the canal, which is west of Port Tennant.

Ordnance Survey Grid Reference: SS 69778 97508

Llechryd Canal

The River Teifi was navigable for centuries from Cardigan to Abercych. In 1765, a tinplate works was established at Doly-coed near Llechryd, and in 1772 a side cut to the river was built between Llechryd and Manordeifi. Llechryd gets its name from the Welsh for 'Slate Ford'. There is dispute about whether the side cut was simply a water channel or a navigation, but there is evidence that it was used by barges carrying timber and iron ore to the works.

Between 1764 and 1770, iron and tinplate works were established on the banks of the Teifi. Materials for the works were brought up the navigable river, and the surrounding woodland provided fuel. The works passed successively through several hands.

In about 1770, the tinplate works was moved across the River Teifi to Pen-y-gored and was bought by Sir Benjamin Hammett, who also purchased the Castell Malgwyn Estate. He built a towpath alongside the river from Pentood Marsh to Llechryd for barges carrying iron, tin and chemicals to his works. The works closed in 1806 and the water course fell into disuse, but the river remained in use as a navigation into the 1990s, transporting quarried stone.

The canal, which has since been infilled, was also used to supply the tinplate works with water. These works were closed in the late eighteenth century.

Ordnance Survey Grid Reference: SN 21295 44284

Mackworth's Canal

Mackworth's Canal lay entirely within Glamorganshire. It was a 300-yard (274.3-metre) tidal cut from a pill on the River Neath about 1 mile (1.6 kilometres) below the town of Neath. It was built to serve Sir Humphrey Mackworth's lead and copper works.

Built between 1695 and 1700, there is some doubt about whether it can be classed as a canal. It certainly began as a canalised tidal pill from the River Neath toward Sir Humphrey Mackworth's works at Melyncryddan. The tidal cut was 18–20 feet (5.4–6.0 metres) wide and capable of taking vessels of up to 100 tons (101.6 tonnes) burden. Soon after the cut was made, a set of heavy wooden gates was added, converting part of the cut into a dock. An early wooden-railed tramroad ran from the dock into the lead and copper works.

By 1720, as shown on O'Connor's map, the cut and dock were not in use. On the map, in place of the dock, is a wharf on the riverside and a tramroad into the works. William Jones' map of 1766 shows the tidal cut marked as 'Melincryddan Pill' and the dock described as 'An old cut for unloading of Oars'.

The Mackworths lived at Gnoll House, which stood on a terrace cut into the side of the hill overlooking the town of Neath. The house had been built for Thomas Evans in 1666. It was successively extended, first by Humphrey Mackworth and then the two Herbert Mackworths, father and son, in the eighteenth century.

Houses were built in the area for the workers at Mackworth's coal mines and 'battery mills' for rolling brass. The canal itself was disused by about 1720.

Ordnance Survey Grid Reference: SS 75613 97526

Penclawdd Canal

The Act of Parliament of the Penclawdd Canal was passed on 21 May 1811. It authorised the construction of a canal – or a railway – from the village of Penclawdd, Glamorganshire, to a point at, or near, Pont Llewydda (Llewitha Bridge), together with a number of branches. Its authorised capital was £12,000 in shares of £100 each, with a further £8,000 to be raised by mortgage if required. The proprietors included Henry Charles, the 6th Duke of Beaufort. The surveyors were Edward Martin from Morriston, Swansea, who was the duke's mining agent, and David Davies of Crickhowell, who estimated the cost for the canal as £9,934. The canal was subsequently built within the estimate.

It was just over 3.5 miles (5.6 kilometres) long and ran from a new dock at Penclawdd, along the coast to Pont-cob, then inland, closely following the course of the River Llan to a terminal basin near Ystrad Isaf. Joseph Priestley recorded that it ran 'in a crooked course from west to east, to the mines not far from the Paper Mill Lands'. There were two locks near what is now the village of Gowerton. The line of the canal was continued by a tramroad towards Llewitha Bridge, and a branch tramroad was built southward from the terminal basin to collieries near the present-day village of Waunarlwydd. None of the other authorised branches was built.

According to the Ship News columns of *The Cambrian* newspaper, between 24 May 1806 and 8 July 1808, twenty-two vessels brought copper ore to Penclawdd from the Cornish ports of Truro, Portreath and Hayle, and six unloaded bricks and clay from Chester. Walter Davies, writing in 1815, implied that the copperworks benefited from the canal, but others seem to suggest that the port only existed to service the works.

According to newspaper advertisements of the time, the canal and dock were open by May 1814. In *The Cambrian* for 28 May 1814, for example, there was an advertisement that read,

> To be Sold Coals, of superior excellence, possessing high bituminous and binding qualities, and durable, calculated for Culinary purposes, Smith's work etc., and for the Irish and Foreign markets, will be ready for shipping on the 13th day of June next, in the New Dock, at the extremity of the Penclawdd Canal, from a New Colliery now opened by Lockwood, Morris, and Leyson, at Wain Arglwydd, known by the name of Poor Man's Coal.

The partnership of Lockwood, Morris & Leyson not only owned coal pits around Waunarlwydd, which the canal served, but also seem to have been the only traders on the canal. The main, or possibly the only, purpose of the canal was to enable the export of coal from Waunarlwydd, but the venture failed before long.

Obviously, the Penclawdd canal was a scheme for the convenient, cheap transport of coal, the same good quality bituminous coal that was extensively mined along the ridges and troughs running westward from Swansea to Llanmorlais. The new quay probably made use of the earlier basin which must have existed at Penclawdd. In the first twenty years of the nineteenth century, the main stream of the Llwchwr River swept in a wide arc below Loughor, close to the edge of the marsh, which gave easy access to Penclawdd. For the same reason, a copperworks had been built at Penclawdd. Writing in 1833, Lewis said, 'In this village were formerly extensive copperworks, belonging to the Cheadle Copper Company, but they are now neglected, that Company possessing numerous others in more convenient situations.' According to the Ship News columns of between 24 May 1806 and 8 July 1808, twenty-two vessels brought copper ore to Penclawdd from Cornish ports (Truro, Portreath and Hayle), and six carried bricks and clay from Chester. The port existed solely as a means of servicing the works.

The canal company had regular meetings until 1818, but after that the canal seems not to have been managed in any way, and it gradually became derelict. On the Ordnance Survey map of 1830 (surveyed in the 1820s), the line of the canal is shown to be complete and seemingly in good order. Similarly, Denham's chart of the Burry estuary of 1830 indicates that it was still in existence. But certainly by 1860 the canal was largely derelict, except for the dock at Penclawdd, which continued in use until the 1870s, when much of it was infilled for the building of a railway.

The remains of the canal today include: the substantial earthworks of the dock (which still holds some water in winter), a length of towpath leading to what is now an ornamental pond on the line of the canal near Berthlwyd, a culvert under the line also near Berthlwyd, a heavily overgrown length near Pont-cob (which includes the abutments of a collapsed overbridge), the chamber of the lower lock near Gowerton, and a partly collapsed bridge over the tail of the lock. During the Second World War, the lock chamber was roofed over and used as an air-raid shelter for the workers of the Fairwood tinplate works – perhaps the only such use of a lock anywhere in the British Isles. The footings of an aqueduct over the Gors-fawr brook, near Gowerton, also still exist, as does the stone arch of an overflow culvert. The line of the extension tramroad can be followed from the northern bank of the River Llan at Ystrad Isaf to Llewitha Bridge, including the abutment of a substantial bridge over the river and an occasional stone sleeper block.

Lewis wrote, in *Samuel Lewis' Topographical Dictionary of Wales*, in 1833: 'In this village were formerly extensive copperworks … but they are now neglected, that Company possessing numerous others in more convenient situations.' Again, according to the Ship News columns, between 24 May 1806 and 8 July 1808, twenty-two vessels brought copper ore to Penclawdd from Cornish ports (Truro, Portreath and Hayle), and six carried bricks and clay from Chester.

Ordnance Survey Grid Reference: SS 54728 95773

Penrhiwtyn Canal

Little is known of the Penrhiwtyn Canal in Glamorganshire, except that it was another of the works carried out by Lord Vernon in around 1792 and cost £600 to build. Penrhiwtyn itself is now a suburb of Neath. The canal, which was almost 1.5 miles (2.4 kilometres) long, became part of the Neath Canal Extension in 1798. It is shown on an old map dated 1791 (among the Eaglesbush Estate papers) running from Cwrt Sart Pill to the ironworks belonging to Raby & Company at Penrhiwtyn. The Evans family who owned the Eaglesbush Estate were related by marriage to the Mackworths of Gnoll, both families being descendants of Edward Evans.

Herbert Evans of Eaglesbush, like Lord Vernon, was a landowner and colliery owner, and the owner of the recently-constructed Penrhiwtyn ironworks. Alexander Raby also had an interest in improving the means of transport in the area. Between the River Neath and Penrhiwtyn was a salt marsh criss-crossed by a network of 'reens', or drainage ditches (from the Welsh *rhewyn* or *rhewin*).

The Penrhiwtyn ironworks were built at the edge of the marsh in 1792, at the same time that Herbert Evans was developing the Eaglesbush colliery on higher ground overlooking the ironworks. Herbert Evans supplied the ironworks with coal, using an inclined plane which ended at the iron furnaces.

Marked on the 1791 map is 'Intended Canal for a Float from CwtSart Pill to the furnace, 12 feet [3.6 metres] wide at Top about 1,000 yards [914.4 metres] in length.' Alongside Cwrt Sart Pill is the comment, 'This pill to be deepened and widened.' The proposed canal is shown to finish at Lord Vernon's Wharfs on the CwrtSart Pill, which, before the building of the canal, had been used for transshipping coal to sea-going ships, brought by a tramroad from Evans' colliery to the wharf. Also on the map, across CwrtSart Pill is marked, 'A bay with sluices to scour the pill below.'

Although the canal is marked as a 'float' at this point, there is evidence that points to it being an inland navigable cut. This would mean that the canal was tidal, in which case the varying tidal levels would have creating difficulties with the loading and unloading of barges moored against the wharfs.

Also on the map, the words 'Proposed Extension of the Neath Canal to Giants Grave', written across the CwrtSart Pill, indicate that an extension of the Neath Canal was being considered, to give better loading and unloading facilities at the terminal basin. On the engineer Thomas Dadford's 1797 map of the Neath Canal extension to Giants Grave, the route of the extension is directly over the line of the Penrhiwtyn Canal. It also crosses the CwrtSart Pill on an aqueduct at the same location as that shown on the earlier map of the Penrhiwtyn Canal.

Ordnance Survey Grid Reference: SS 74444 96235

Pen-y-fan (Wern) Canal

Mynydd (mountain) Pen-y-Fan is the highest peak in the Brecon Beacons at just under 3,000 feet (914.4 metres). Pen-y-Fan Pond, in its shadow, is one of the few remaining canal feeder reservoirs in Wales. The reservoir, which now lies within the Penallta country park, is a scheduled ancient monument. The Pen-y-Fan (or Wern) Canal ran for a mile (1.6 kilometres) from near the site of the Copperhouse Dock to Capel Als, Llanelly.

Sir Henry Protheroe purchased the Penivan (Pen-y-Fan) colliery, situated on the eastern slopes of Mynydd Pen-y-Fern near Aberbeeg, in about 1810. He owned it until 1838, when Aaron Crossfield and a partner took it over, the Royal Commission on Mines was told in 1842. Thomas Powell was the owner from 1854–1864, when George Elliot & Company bought it, only to sell it the following year to the Ebbw Vale Company. By 1868 it had changed hands again and Roger Lewis was the owner.

The coal was brought out from the workings along horse-drawn tramroads. A reference made in 1750, to a canal from the colliery, is borne out by the fact that the tunnel into the mine is partly lined with masonry, and that there is a constant and voluminous outflow of water, sufficient to supply the canal.

According to the Inspector of Mines in 1896, twenty-eight men were employed at Pen-y-Fan. By 1908, forty-eight men were working there.

Ordnance Survey Grid Reference: SS 69535 96776

Rhyddings (Redding) Canal

Very little is known about the Rhyddings Canal, near Bryncoch in Neath, which was originally known as the Redding Canal. George Tennant, a London solicitor and entrepreneur, moved into the area and bought the Rhyddings Estate from the Earl of Jersey in 1814. He decided to reopen the Redding colliery on the estate, which was located near the Duffryn Arms pub at Bryncoch, and was worked by a level.

Reference to the colliery being in existence in 1767 was made in the 'Report on the Petition of Owners of Collieries' carried out by a House of Commons committee in 1810. The report also said that the colliery was drained by means of a steam engine and a water wheel in the drainage leat from the colliery.

Coal had been produced below Mynydd (Mount) March Hywel for some time, and the canal had, probably, originally been no more than the leat for water from the mine workings. It followed the contours of the mountain and, when George Tennant bought the estate, it was probably he who extended and widened it and introduced wooden tub boats to transport the coal from the collieries to a point near Redding Farm, which is now known as Rhyddings Yard Farm, situated in the area of land known as the Rhyddings. From there, the tub boats were transferred to an inclined tramway which connected to the River Neath near the old river bridge in Neath. However, the tramway crossed part of the Ynysygerwn Estate without permission from the owner, Mr Lewis Weston Dillwyn. Eventually it was diverted to pass down the side of Cadoxton churchyard, to a wharf on the Tennant Canal. Trespassing on Dillwyn's land would later have serious repercussions when George Tennant began to build his canal from Red Jacket to Aberdulais.

It is not known how long the Rhyddings Canal remained in use, but today the towpath is still used as a footpath and a stream flows along part of the canal, which is still owned by the Coombe Tennant Estate. Workers from the estate occasionally clean out the water course to prevent silting up and flooding.

Ordnance Survey Grid Reference: SS 75000 98527

Rhyd-y-defaid Mining Canal

The Rhyd-y-defaid Canal was located at Blackpill in the Clyne valley, which is now a suburb of Swansea. Only a little of this canal, which was probably built in 1840, remains. It was used to supply coal to Swansea copper smelters. In its nineteenth-century industrial heyday, Swansea was a major centre for the copper industry and earned itself the nickname 'Copperopolis'.

The first Welsh copper smelting works had been established at Aberdulais in 1584. It was in the Neath valley, in the late seventeenth century, that copper smelting, refining and working first became a commercial concern in Wales. The valley's waterfalls provided the power for two copper mills near Neath Abbey, the first of which opened in 1694. These works supplied the copper and brass manufacturing industries in Bristol and London.

Swansea's combination of being a port with local collieries, together with trading links with the English West Country, meant that it was the obvious location to build works for smelting copper. Copper mining is probably the oldest known mining activity in Wales. The Welsh copper industry, together with Cornwall, had a virtual monopoly in the global trade in the late eighteenth century.

Smelters were operating by 1720, and their number grew. With the growing demand for coal, more collieries were opened, everywhere from the north-east of the Gower peninsula to Clyne and Llangyf. Smelters were opened mostly along the Tawe valley and flourished. To reach the smelting works, the coal was carried, probably by tub boats, from the colliery along the canal to Blackpill, and then to Swansea by way of the Mumbles Railway.

Thomas Williams, a lawyer-entrepreneur from Llanidan, Anglesey, a friend of John 'Iron Mad' Wilkinson, formed the Parys Mine Company in 1778 to exploit the copper ore seams that had been discovered on the island. The Parys company became, in the 1780s and 1790s, the world's most productive copper mine, eclipsing the Cornish mines, producing about 44,000 tons (44,706 tonnes) a year at its peak.

The household coal carried on the canal was mined at the Rhyd-y-defaid Colliery at Killay, which was owned by Phillip Richards of Swansea.

Ordnance Survey Grid Reference: SS 62042 90760

Swansea Canal

In 1790 Edward Martin carried out a survey for William Padley, a Swansea merchant, for a canal up the Tawe valley. A second survey was carried out in 1793 by Thomas Sheasby Snr. The Swansea Canal Act was passed by Parliament in May 1794, authorising the proprietors to raise £60,000, by the sale of shares, to build a canal from Swansea to Hen Neuadd, near Abercrave.

It was decided to construct the canal using direct labour rather than by employing a contractor. Charles Roberts, who had been dismissed as the engineer of the Manchester, Bolton & Bury Canal for not having 'acted in the Execution of his Duty with proper discretion and Economy' was appointed engineer, but was dismissed after nine months. Thomas Sheasby Snr was then appointed, assisted, at first, by his son. Later, Sheasby Snr continued alone, although he was in dispute with the Neath Canal at the time.

The plans were opposed by the Duke of Beaufort and others, who wanted the canal to terminate further up the river near Landore and Morriston, where they already had wharfs. Swansea Corporation favoured the route into Swansea, and offered to contribute towards the cost, at which point the duke, his son, the Marquess of Worcester, and the duke's agent withdrew their subscriptions for shares. However, this action stimulated others to subscribe, and £52,000 was raised almost immediately.

Ultimately, a compromise was reached, with the canal terminating in Swansea, but the duke constructing just under 1.5 miles (2.3 kilometres) of canal from Nant Rhydyfiliast to Nant Felin, on which he was allowed to charge tolls – provided they did not exceed the tolls charged by the canal company for use of the rest of the canal. The duke's section, which was called the Trewyddfa Canal, was part of the main line.

The Swansea Canal was 16.5 miles (26.5 kilometres) long and had a rise of 300 feet (91.4 metres) through thirty-seven locks. It opened to Godre'rgraig by 1796 and was fully open by October 1798. The company's locks had been built to take boats carrying 25 tons (25.4 tonnes) of 69 feet by 7 feet 6 inches (21 metres by 2.2 metres), but those on the duke's section were only 65 feet (20 metres) long, which imposed restrictions on boats using the whole or large parts of the canal. There were no tunnels, but there were five aqueducts to carry the canal across major tributaries of the Afon Tawe, at Clydach, Pontardawe, Ynysmeudwy, Ystalyfera, and Cwmgiedd; the largest of them was across the Twrch at Ystalyfera. Unusually, the final cost of building the canal was well within budget, costing £51,602 up to mid-1798.

Once the canal was open, some members of the committee called for it to be extended from Swansea to Oystermouth to carry limestone from Mumbles and coal from the Clyne valley into Swansea. The extension was opposed by other members of the committee, but eventually a compromise was reached through the good offices of Edward Martin, whereby a tramroad was to be built. This received its Act of Parliament in 1804.

The Swansea Canal was served by other tramroads from the many collieries in the Tawe valley, one of the largest of which was the Brecon Forest tramroad, completed in 1834. In addition, there were private branch canals from the Ynyscedwyn ironworks, Ystalyfera ironworks, and collieries at Cilybebyll and Pontardawe. Eventually, a branch was built along the north dock in Swansea in 1852.

The canal was different to every other canal in the British Isles, in that nearly all of its shareholders were owners of the industries along its bank. They therefore had the advantage of being able to use the flow of water along the canal as a source of power for their works. During the life of the canal, forty-two leats were built along its route for this purpose.

In 1804, 54,235 tons (55,105.3 tonnes) of coal and culm were carried, and the company's profits were enough for a dividend of 3 per cent to be paid. Receipts and dividends rose steadily, reaching £10,522 and 14 per cent respectively in 1840, while in 1860 they were £13,800 and 18 per cent. There are few surviving records of how much cargo was carried, but estimates based on the amount of coal and culm known to have been shipped from Swansea docks suggest that around 386,000 tons (392,194.1 tonnes) could have been carried in 1839.

The canal prospered at first, but the opening of the Swansea Vale Railway in 1852 provided direct competition, and by the 1860s receipts had fallen despite a reduction in carrying rates in 1864. It was not only the railway that created difficulties for the canal, but the growth in the use of steel by industry in preference to iron as well. The last of the ironworks in the Tawe valley had closed by 1886, and the coal trade had already been lost to the railways. When the Neath & Brecon Railway arrived in 1864, following the Swansea Vale Railway, it further affected trade on the canal.

The harbour facilities at Swansea were upgraded in 1852, when the River Tawe was diverted into a new channel to the east and the original channel, which included the trans-shipment wharfs, became a floating harbour. A lock was constructed to give boats from the canal direct access to the half-tide basin above the north dock, and a loop of the canal was constructed along the edge of the new harbour.

In 1873, the canal was bought out by the Great Western Railway (GWR), which had absorbed the Swansea Vale and Neath & Brecon railways. Ownership of the canal enabled the GWR to compete with the rival Midland Railway until, in 1895, the canal made its first working loss. Nevertheless, carrying continued until 1931. The last narrowboat built for use on the canal was in 1918, when *Grace Darling* was built at the Godre'r Graig boatyard.

The canal was abandoned piecemeal, first under the Great Western Railway Acts of 1928 and 1931, then under the British Transport Commission Acts of 1949 and 1957. The closure of the upper end came under the Brecknockshire County Council Act of 1946, the 2-mile (3.2-kilometre) section between Ynysmeudw and Ystalyfera by warrant in 1961, and the remainder by the British Transport Commission Act of 1962.

Much of the canal has been infilled over the past fifty years. Today, just 5 miles (8 kilometres) of the canal remains in water, from Clydach to Pontardawe. The Swansea Canal Society is actively involved in the restoration of the length that has not been infilled.

Ordnance Survey Grid Reference: SS 75000 98527

Swansea Canal. Image used by permission of Jeff Griffiths.

Tremadoc Canal

Tremadog (which was formerly called Tremadoc), a village on the outskirts of Porthmadog, in Gwynedd, north Wales, was a planned settlement founded by William Alexander Madocks, who bought the land on which it stands in 1798. The land had been recovered from the Traeth Mawr by the building of an embankment in 1800. The centre of Tremadog was complete by 1811 and has changed little since then.

Madocks enlarged a drainage ditch to the River Glaslyn to form a canal (known locally as Y Cyt) 1.5 miles (2.4 kilometres) long with no locks, named after the village. It was opened around 1815 and was used for thirty-five years to carry copper ore from a local mine before it was replaced by a tramroad. The canal was provided at Madocks' expense to link Llyn Bach and Porthmadog Harbour to a basin at Tremadoc, where he lived in a house called Tan yr Allt. The house was later home to the poet Percy Bysshe Shelley, and then to several generations of a family of local quarry owners.

In 1855/56, the Gorseddau slate quarry, located at the head of the valley of Cwmystradllyn, saw huge expansion, with nine levels, a slate mill, and a tramway to Portmadoc called the Gorseddau Tramway.

Part of the tramway's route is still visible alongside the restored narrow gauge Welsh Highland Railway's car park in Porthmadog. It resembles nothing more than an overgrown hedge-lined ditch, crossed by a rickety wooden footbridge.

Ordnance Survey Grid Reference: SH 56216 40071

The bridge that carried the Croesor Tramway and later the Welsh Highland Railway over the Tremadoc Canal. Image used by permission of Roger Marks.

Trewyddfa (Morris's) Canal

Morris's Canal in Glamorganshire (later incorporated in the Trewyddfa Canal) was built by David Morris in 1783/84 as an extension of the Clyndu Level underground mining canal. It ran from the Forest Copper Works to Landore, and used boats 29 feet by 3 feet (8.8 metres by 0.9 metres), the same size as those on the Clyndu Level. Coal from the level was used by the Forest Copper Works and later by three smelting works along the River Tawe.

After the 1.25-mile (2-kilometre) canal was enlarged in 1790 by the Duke of Beaufort, it became known as the Trewyddafa Canal.

When the Swansea Canal was promoted, its committee said, having taken legal advice, that '… this canal [the Trewyddfa] has been made over a great part of the Manor of Trewyddfa without the consent of the Holders and Homagers and without the sanction of Parliament'. When the Swansea Canal Act was passed in 1794, the Trewyddfa Canal became part of that canal, although the Duke of Beaufort retained ownership of the Trewyddfa Canal until 1873.

Its existence was acknowledged in the Swansea Canal Company's Act, and traffic was allowed to pass through the length of the duke's section of the Swansea Canal without paying tolls to the Swansea Company, while the duke was allowed to charge his tolls on through traffic.

Infilling of much of the Swansea Canal has included the Trewyddfa Canal, so there are now no visible remains.

Ordnance Survey Grid Reference: SS 66082 95649

Maps

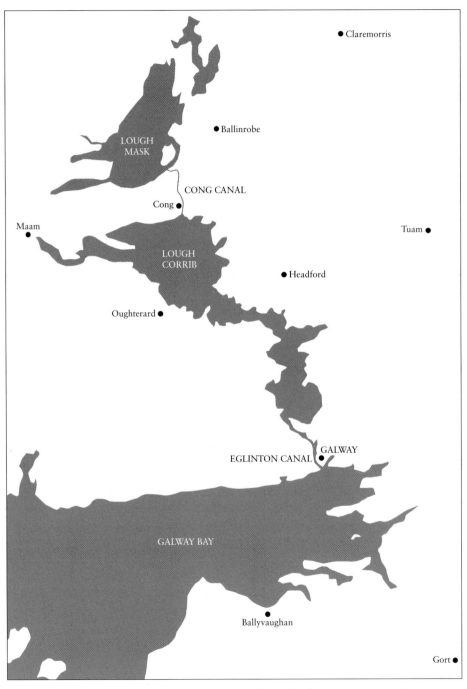

A map showing the Cong and Eglinton Canals. (Map drawn by Thomas Bohm, User Design, Illustration and Typesetting)

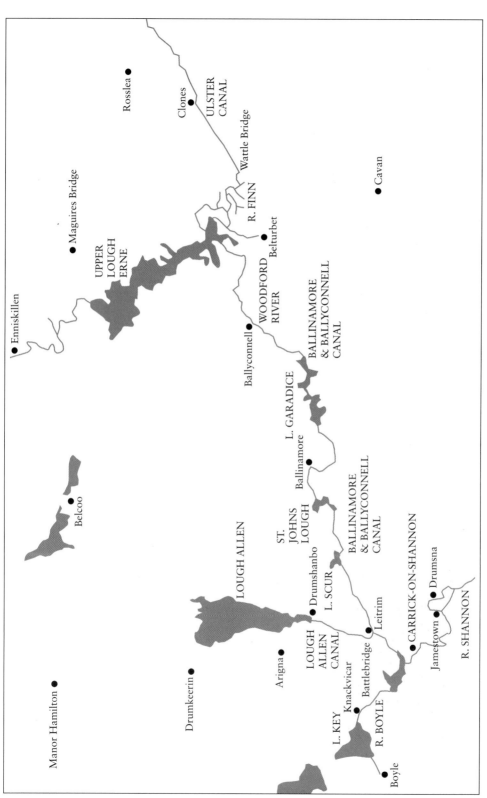

A map showing the Ulster Canal. (Map drawn by Thomas Bohm, User Design, Illustration and Typesetting)

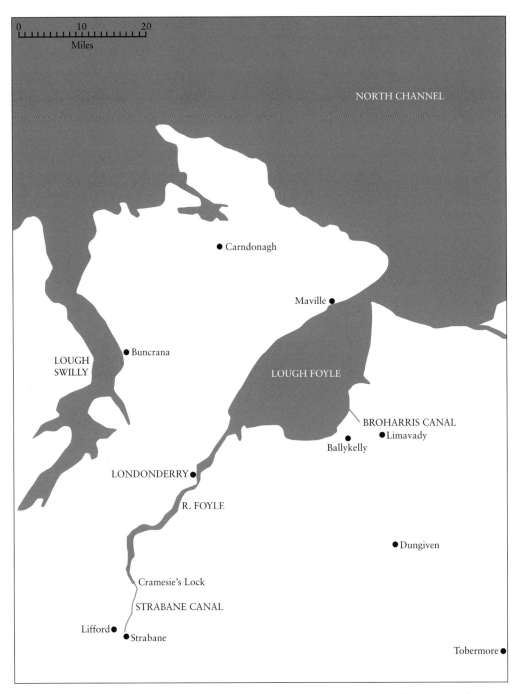

A map showing the Broharris Canal. (Map drawn by Thomas Bohm, User Design, Illustration and Typesetting)

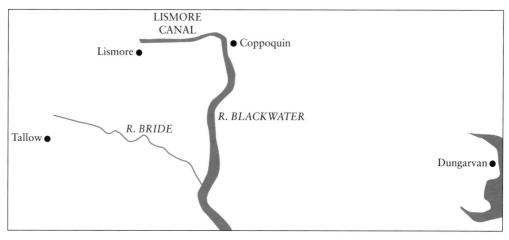

A map showing the Lismore Canal. (Map drawn by Thomas Bohm, User Design, Illustration and Typesetting)

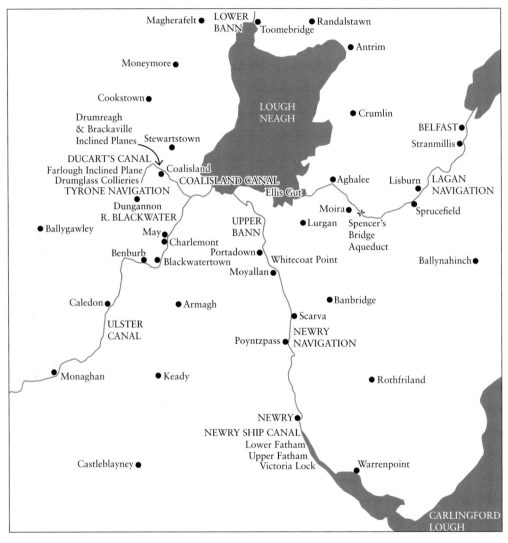

A map showing the Coalisland, Newry and Ulster Canals. (Map drawn by Thomas Bohm, User Design, Illustration and Typesetting)

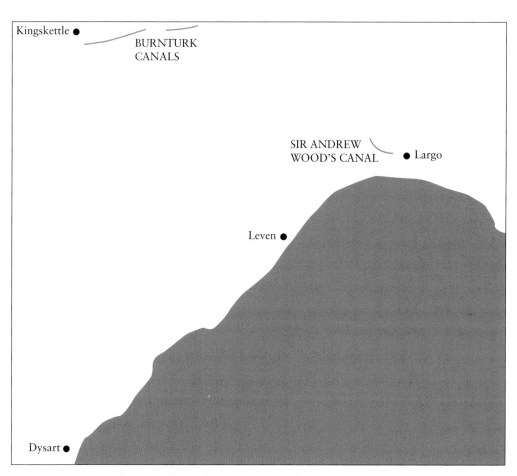

A map showing Sir Andrew Wood's Canal. (Map drawn by Thomas Bohm, User Design, Illustration and Typesetting)

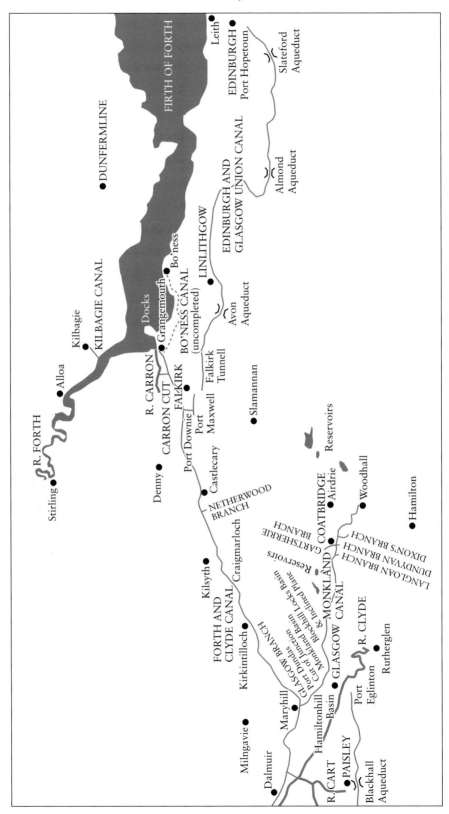

A map showing the Bo'ness, Carron, Kilbagie and Monkland Canals. (Map drawn by Thomas Bohm, User Design, Illustration and Typesetting)

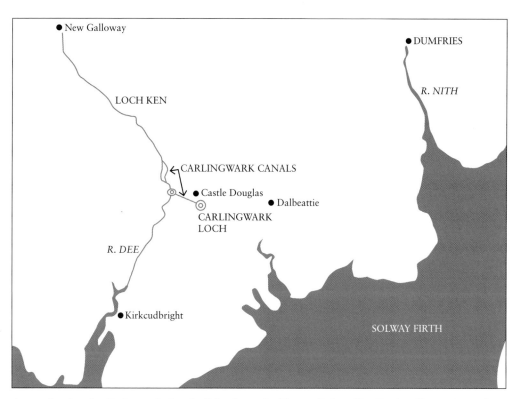

A map showing the Carlingwark Canals. (Map drawn by Thomas Bohm, User Design, Illustration and Typesetting)

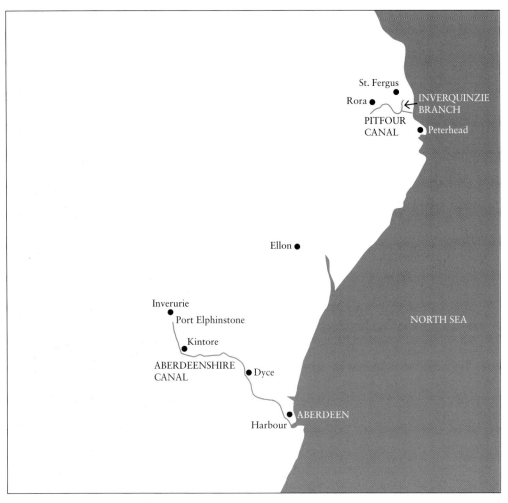

A map showing the Aberdeenshire and Pitfour Canals. (Map drawn by Thomas Bohm, User Design, Illustration and Typesetting)

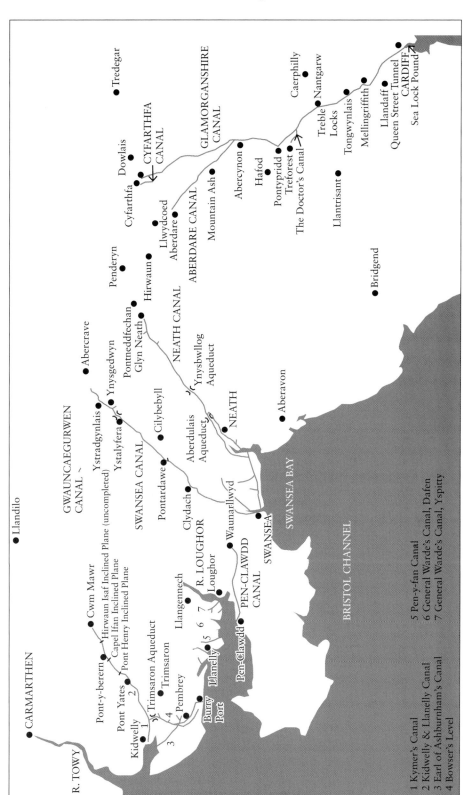

A map showing Bowser's Level, Cyfarthfa, Earl of Ashburnham's, General Warde's (Dafen and Yspitty), Glamorganshire, Gwaun-Cae-Gurwen, Kidwelly & Llanelly, Kymer's, Penclawdd, Pen-y-fan and Swansea Canals. (Map drawn by Thomas Bohm, User Design, Illustration and Typesetting)

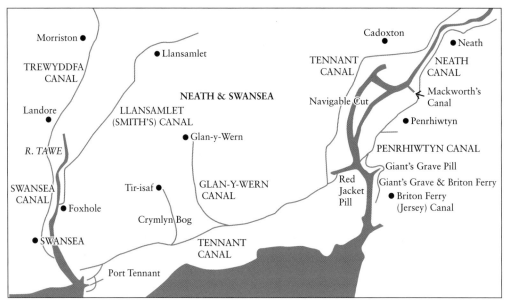

A map showing the Giant's Grave & Briton Ferry, Glan-y-Wern, Llansamlet, Mackworth's, Penrhiwtyn, Swansea and Trewyddfa Canals. (Map drawn by Thomas Bohm, User Design, Illustration and Typesetting)

Bibliography

General
Paget-Tomlinson, Edward W., *The Complete Book Of Canal & River Navigations* (Waine Research Publications).

Ireland

Athlone Canal
Cumberlidge, Jane, *The Inland Waterways of Ireland* (Imray Laurie Norie & Wilson).
Goggin, Brian J., 'Athlone Canal', *Irish Waterways History* (http://irishwaterwayshistory. com/abandoned-or-little-used-irish-waterways/the-middle-and-upper-shannon/athlone-canal/).
Hamond, Fred, 'Industrial Heritage of the Shannon Waterway' (www.heritagecouncil.ie).

Ballinasloe Canal
'Grand Canal Ballinasloe Branch', *Industrial Heritage Ireland* (http://industrialheritageireland. info/TikiWiki/tiki-index.php?page=Grand+Canal+-+Ballinasloe+Branch).

Ballycuirke Canal
Cumberlidge, Jane, *The Inland Waterways of Ireland* (Imray Laurie Norie & Wilson).
Goggin, Brian J., 'Ballycuirke Canal', *Irish Waterways History* (http://irishwaterwayshistory. com/abandoned-or-little-used-irish-waterways/waterways-of-the-west/the-ballycuirke-canal/).

Bannagher Boat Drain
Delany, Ruth, *By Shannon Shores* (Gill & McMillan).
Delaney, Ruth, *The Shannon Navigation* (Lilliput Press).
Goggin, Brian J., 'Bannagher Boat Drain', *Irish Waterways History* (http:// irishwaterwayshistory.com/abandoned-or-little-used-irish-waterways/shannon-south/ irelands-shortest-canal/).

Bridgetown Canal
Cumberlidge, Jane, *The Inland Waterways of Ireland* (Imray Laurie Norie & Wilson).
'Bridgetown Canal History', *Inland Waterways Association of Ireland, Slaney Branch* (http://slaney.iwai.ie/btown_hist.htm).

Broadstone Branch (Royal Canal)
Cumberlidge, Jane, *The Inland Waterways of Ireland* (Imray Laurie Norie & Wilson).
Delaney, Ruth and Ian Bath, *Ireland's Royal Canal 1789-2009* (Lilliput Press).

Goggin, Brian J., 'The Broadstone Line of the Royal Canal' (http://irishwaterwayshistory.com/
abandoned-or-little-used-irish-waterways/waterways-in-dublin/the-broadstone-line-of-the-
royal-canal/).

Broharris Canal
Cumberlidge, Jane, *The Inland Waterways of Ireland* (Imray Laurie Norie & Wilson).
Delany, Ruth, *Ireland's Inland Waterways* (Appletree Press).
'Strabane and Broharris Canals', *Ireland's Eye* (http://www.irelandseye.com/irish/travel/water/
strabane.shtm).

Coalisland (Tyrone) Canal and Ducart's Canal
'Davis Ducart', *Dictionary of Irish Architects* (http://www.dia.ie/architects/view/1660).
McGough, Hugh, 'Coal Mining and Canal Building in County Tyrone in the Eighteenth and
Nineteenth Centuries' (http://magoo.com/hugh/tyronecoal.html).
Ransom, P. J. G., 'Ireland', *The Archaeology of Canals* (World's Work).
'Coalisland Canal', *Unlocking Lough Neagh's Heritage* (http://www.loughneaghheritage.com/
Navigation-Transport/canals-%281%29.aspx).

Cong Canal
Goggin, Brian J., 'The Cong Canal', *Irish Waterways History* (http://irishwaterwayshistory.com/
abandoned-or-little-used-irish-waterways/waterways-of-the-west/the-cong-canal/).
'Cong Canal', *Ireland's Lake District Heritage* (http://www.lakedistrictheritage.ie/Cong/canal.html).
Semple, Maurice, *By The Corribside* (Maurice Semple).

Eglinton Canal
'The construction of the Eglinton Canal', *Galway Independent* (http://galwayindependent.
com/20130227/news/the-construction-of-the-eglinton-canal-S7774.html).
'The opening of the Eglinton Canal', *Galway Independent* (http://galwayindependent.
com/20130315/news/the-opening-of-the-eglinton-canal-S029.html).
Goggin, Brian J., 'The Eglinton Canal in Galway', *Irish Waterways History* (http://
irishwaterwayshistory.com/abandoned-or-little-used-irish-waterways/waterways-of-
the-west/the-eglinton-canal-in-galway/).
National Inventory of Architectural Heritage, 'Eglinton Canal Basin, Claddagh Quay, Galway,
County Galway' (http://www.buildingsofireland.ie/).

Finnery River Navigation
Cumberlidge, Jane, *The Inland Waterways of Ireland* (Imray Laurie Norie & Wilson).
Goggin, Brian J., 'The Finnery River Navigation', *Irish Waterways History* (http://
irishwaterwayshistory.com/abandoned-or-little-used-irish-waterways/midlands-turf-
waterways/the-finnery-river-navigation/).

Grand Canal Docks (Abandoned Line)
Delaney, Ruth, *Ireland's Inland Waterways* (Appletree Press).
Goggin, Brian J., 'The abandoned Main Line of the Grand Canal', *Irish Waterways History*
(http://irishwaterwayshistory.com/abandoned-or-little-used-irish-waterways/waterways-
in-dublin/the-abandoned-main-line-of-the-grand-canal-1/).

John's Canal, Castleconnell

Cumberlidge, Jane, *The Inland Waterways of Ireland* (Imray Laurie Norie & Wilson).

Goggin Brian J., 'John's Canal, Castle Connell', *Irish Waterways History* (http://irishwaterwayshistory. com/abandoned-or-little-used-irish-waterways/midlands-turf-waterways/johns-canal-castleconnell/).

Hannan, Kevin, 'Castleconnell', *Limerick City Council* (http://www.limerickcity.ie/media/ Media,3963,en.pdf).

Lismore Canal

Delaney, Ruth, *Ireland's Inland Waterways*, (Appletree Press).

Goggin, Brian J., 'The Bride, the Munster Blackwater and the Lismore Canal', *Irish Waterways History* (http://irishwaterwayshistory.com/abandoned-or-little-used-irish-waterways/waterways-of-cork-and-kerry/the-bride-the-munster-blackwater-and-the-lismore-canal/).

Lynch's Canal

Madden, Marie, 'The Construction of the Eglinton Canal', *Galway Independent* (http:// galwayindependent.com/20130227/news/the-construction-of-the-eglinton-canal-S7774.html).

Mountmellick Line (Grand Canal)

Cumberlidge, Jane, *The Inland Waterways of Ireland*, (Imray Laurie Norie & Wilson).

Delaney, Ruth, *Ireland's Inland Waterways* (Appletree Press).

Goggin, Brian J., 'The Mountmellick Line of the Grand Canal', *Irish Waterways History* (http:// irishwaterwayshistory.com/abandoned-or-little-used-irish-waterways/the-grand-canal/the-mountmellick-line-of-the-grand-canal/).

Ransom, P. J. G., *The Archaeology of Canals* (World's Work).

Newry Canal

Cumberlidge, Jane, *The Inland Waterways of Ireland* (Imray Laurie Norie & Wilson).

Delaney, Ruth, *Ireland's Inland Waterways* (Appletree Press).

McCutcheon, W. A., *The Canals of the North of Ireland* (David & Charles).

Park Canal

Cumberlidge, Jane, *The Inland Waterways of Ireland* (Imray Laurie Norie & Wilson).

Delaney, Ruth, *Ireland's Inland Waterways* (Appletree Press).

Goggin, Brian J., 'The Park Canal, why it should not be restored', *Irish Waterways History* (http://irishwaterwayshistory.com/rants/the-park-canal-why-it-should-not-be-restored/).

Plassey–Errina and Killoloe Canals

Goggin, Brian J., 'Plassey-Errina Canal', *Irish Waterways History* (http://irishwaterwayshistory. com/tag/plassey-errina-canal/).

'Limerick - Navigation history', *Inland Waterways News*, Inland Waterways Association Ireland (http://iwn.iwai.ie/v28i2/limerickoldcanal.PDF).

Wikipedia, 'Ardnacrusha Power Plant' (http://en.wikipedia.org/wiki/Ardnacrusha_power_plant).

Rockville Navigation

Goggin, Brian J., 'The Rockville Navigation', *Irish Waterways History* (http://irishwaterwayshistory.com/abandoned-or-little-used-irish-waterways/midlands-turf-waterways/the-rockville-navigation/).

Rockingham Canal

Goggin, Brian J., 'Rockingham', *Irish Waterways History* (http://irishwaterwayshistory.com/abandoned-or-little-used-irish-waterways/midlands-turf-waterways/rockingham/).
Lough Key Forest and Activity Park, 'History of Rockingham and Lough Key' (http://www.loughkey.ie/park-and-estate/history.html).

Roscrea Canals

Goggin, Brian J., 'The Roscrea Canals', *Irish Waterways History* (http://irishwaterwayshistory.com/abandoned-or-little-used-irish-waterways/midlands-turf-waterways/the-roscrea-canals/).
'Roscrea Whiskey', *Roscrea Online* (http://www.roscreaonline.ie/content.asp?section=1057).

Ulster Canal

Cumberlidge, Jane, *The Inland Waterways of Ireland* (Imray Laurie Norie & Wilson).
Delany, Ruth, *Ireland's Inland Waterways* (Appletree Press).
McCutcheon, W. A., *The Canals of the North of Ireland* (David & Charles).
Ransom, P. J. G., *The Archaeology of Canals* (World's Work).

Scotland

Aberdeenshire Canal

'The Aberdeenshire Canal', *James's Canal Pages* (http://www.jamescanalpages.org.uk/aberdeenshire.php).
Lindsay, Jean, *The Canals of Scotland* (David & Charles).
Priestley, Joseph, *Historical Account of the Navigable Rivers, Canals, and Railways of Great Britain* (Longman, Rees, Orme, Brown & Green). Not in copyright (https://archive.org/details/historicalaccou01priegoog).

Borrowstouness (Bo'ness) Canal

Lindsay, Jean, *The Canals of Scotland* (David & Charles)
Priestley, Joseph, *Historical Account of the Navigable Rivers, Canals, and Railways of Great Britain* (Longman, Rees, Orme, Brown & Green). Not in copyright (https://archive.org/details/historicalaccou01priegoog).
Wright, Ken, 'The Borrowstounness Canal Company' (http://www.bo-ness.org.uk/html/history/canal_company.htm).

Black Wood of Rannoch Canals

Briggs, Robert and Jude Mitchell, 'Canadian Forestry Corps' (http://freepages.genealogy.rootsweb.ancestry.com/~jmitchell/cfc39.html).
Fargo, Jim, 'The Black Wood of Rannoch, Roberton's Rant', *The Magazine of the Clan Donnachaidh Society*, Vol. 2, Issue 3, November 2013.
'Black Wood of Rannoch', *Gazetteer for Scotland* (http://www.scottish-places.info/features/featurefirst14383.html).

Campbeltown and Machrihanish Coal Canal
'Coal Mining in Kintyre', *Campbeltown Heritage Centre* (http://www.campbeltownheritagecentre.co.uk/coalmining.php).

Farr, A. D., *The Campbeltown & Machrihanish Light Railway* (The Oakwood Press).

'Campbeltown to Machrihanish', *Kintyre Way* (http://www.kintyreway.com/campbeltown-machrihanish.php).

Lindsay, Jean, *The Canals of Scotland* (David & Charles).

Carlingwark Canals
Lindsay, Jean, *The Canals of Scotland* (David & Charles).

Livingston, Alastair, 'The Rationalised Landscape', *Green Galloway* (http://greengalloway.blogspot.co.uk/2011/10/rationalised-landscape.html).

'Carlingwark Canal', *Lost Canals*, Secret Scotland (http://www.secretscotland.org.uk/index.php/Secrets/LostCanals).

Carron Cut
'Carron Canal', *Lost Canals*, Secret Scotland (http://www.secretscotland.org.uk/index.php/Secrets/LostCanals#Carron_Canal).

Forth & Cart Canal
Cumberlidge, Jane, *Inland Waterways of Great Britain* (Imray Laurie Norie & Wilson).

Lindsay, Jean, *The Canals of Scotland* (David & Charles).

'Forth & Cart Canal', *UK Canals Network* (http://www.ukcanals.net/waterways-of-the-uk/131-forth-a-cart-canal).

Glasgow, Paisley & Ardrossan Canal
Lindsay, Jean, *The Canals of Scotland* (David & Charles).

Priestley, Joseph, *Historical Account of the Navigable Rivers, Canals, and Railways of Great Britain* (Longman, Rees, Orme, Brown & Green) Not in copyright (https://archive.org/details/historicalaccou01priegoog).

Scott, Walter, 'Melancholy Accident', *The Edinburgh Annual Register*, Vol. 3, Part 2 (John Ballantyne).

Inverarnan Canal
'History of the Inverarnan Canal', *Arrochar, Tarbet and Ardlui Heritage* (www.arrocharheritage.com).

Lappin, Graham A., 'The Loch Lomond Steamers' (http://www.valeofleven.org.uk/lochlomondsteamers.html).

Lindsay, Jean, *The Canals of Scotland* (David & Charles).

'Inverarnan Canal', *Lost Canals*, Secret Scotland (http://www.secretscotland.org.uk/index.php/Secrets/LostCanals#Inverarnan_Canal).

Kilbagie Canal
'Kilbagie Mill', *Royal Commission on the Ancient and Historical Monuments of Scotland* (http://canmore.rcahms.gov.uk/en/site/48118/details/kilbagie+mill/).

'Kennetpans and Kilbagie', *The History of Kennetpans* (http://www.kennetpans.info/index.php option=com_content&view=article&id=135&Itemid=234).

Monkland Canal

Ransom, P. J. G., *The Archaeology of Canals* (World's Work).
Russell, Ronald, *Lost Canals & Waterways of Britain* (Sphere).
'Monkland Canal', *Wikipedia* (http://en.wikipedia.org/wiki/Monkland_Canal).

Muirkirk Canal

Lindsay, Jean, *The Canals of Scotland* (David & Charles).
'Muirkirk Canal', *Muirkirk* (http://www.muirkirk.org.uk/heritage/muirkirk-canal.htm).

Pudzeoch Canal

Biddulph, B. and Stuart A. Cameron, 'Guide to 50 Years on the Clyde', *Clydesite Magazine*
    (http://www.clydesite.co.uk/articles/upperriver.asp)
'Pudzeoch Canal', *Secret Scotland* (http://www.secretsscotland.org.uk/index.php/Secrets/
    LostCanals#Pudzeoch_Canal).
'The Pudzeoch', *Wikimapia* (http://wikimapia.org/8433501/The-Pudzeoch).

Sir Andrew Wood's Canal

Bell, Dennis, 'Sir Andrew Wood', *Tour Scotland* (http://www.fife.50megs.com/sir-andrew-wood.
    htm).
'Sir Andrew Wood', *Significant Scots*, Electric Scotland (http://www.electricscotland.com/
    history/other/wood_andrew.htm).
'Upper Largo, Largo Home Farm, Sir Andrew Wood's Canal', *Royal Commission on the Ancient
    and Historical Monuments of Scotland* (http://canmore.rcahms.gov.uk/en/site/32840/digital_
    images/upper+largo+largo+home+farm+sir+andrew+wood+s+canal/).

St Fergus and River Ugie (Pitfour's) Canal

Buchan, James, *Crowded with Genius* (Harper Collins).
Graham, Angus, 'Two Canals in Aberdeenshire', *Proceedings of the Society of Antiquaries of
    Scotland* 100: 176.

Stevenston (Saltcoats) Coal Canal

Graham, Eric J., 'Robert Reid Cunninghame of Seabank House', *Ayrshire Archaeological and
    Natural History Society*.
Hughson, Irene, *The Auchenharvie Colliery* (Richard Stenlake).
Lindsay, Jean, *The Canals of Scotland* (David & Charles).

Wales

Aberdare Canal

'The Aberdare Canal', *Fletcher Family* (http://www.fletcher-family.co.uk/
    the%20aberdare%20canal.html).
Hadfield, Charles, *The Canals of South Wales and the Border* (David & Charles).
Priestley, Joseph, *Historical Account of the Navigable Rivers, Canals, and Railways of Great
    Britain* (Longman, Rees, Orme, Brown & Green) Not in copyright (https://archive.org/
    details/historicalaccou01priegoog).

'Our Past, Aberdare Canal', *Rhondda Cynon Taf Library Service* (http://webapps.rhondda-cynon-taff.gov.uk/heritagetrail/english/cynon/aberdare_canal.html).

Rowson, Stephen and Ian L. Wright, *The Glamorganshire and Aberdare Canals* Vol. 1 (Black Dwarf Publications).

Bowser's Level

Hadfield, Charles, *The Canals of South Wales and the Border* (David & Charles)

'Canals', *Carmarthenshire Family History Society* (http://www.carmarthenshirefhs.info/pembreye.htm).

'Bowser's Level', *Short History of South Wales Canals* (http://www.visitsouthwalescanals.co.uk/history.pdf).

Burry and Loughor Rivers

Hadfield, Charles, *The Canals of South Wales and the Border* (David & Charles)

Lewis, Samuel, *A Topographical Dictionary of Wales* (S. Lewis & Company) (http://books.google.co.uk/books?id=6XwOAAAAQAAJ&printsec=frontcover&source=gbs_ge_summary_r&cad=0#v=onepage&q&f=false).

Cemlyn Canal

'The Vale of Ffestiniog', *Historic Landscape Characterisation*, Cadw (http://www.heneb.co.uk/ffestiniogcharacter/ffeshisteng.html).

Paget-Tomlinson, Edward W., *The Complete Book Of Canal & River Navigations* (Waine Research Publications).

Clyndu Canal

'Clyndu Canal', *Short History of South Wales Canals* (http://www.visitsouthwalescanals.co.uk/history.pdf).

'Welsh History Month: The lower Swansea Valley', *Wales Online* (http://www.walesonline.co.uk/news/wales-news/welsh-history-month-lower-swansea-2032065).

Crymlyn Canal

Cumberlidge, Jane, *Inland Waterways of Great Britain* (Imray Norie Laurie & Wilson).

Hadfield, Charles, *The Canals of South Wales and the Border* (David & Charles).

Priestley, Joseph, *Historical Account of the Navigable Rivers, Canals, and Railways of Great Britain* (Longman, Rees, Orme, Brown & Green) Not in copyright (https://archive.org/details/historicalaccou01priegoog).

Cyfarthfa Canal

'River Taff and Railway Corridor', *Countryside Council for Wales* (http://www.ggat.org.uk/cadw/historic_landscape/Merthyr%20Tydfil/English/Merthyr_014.htm).

'Crawshay Brothers (Cyfarthfa) Ltd', *Durham Mining Museum* (http://www.dmm.org.uk/company/c1044.htm).

Russell, Ronald, *Lost Canals of England and Wales* (David & Charles).

Doctor's Canal

Davies, John and Nigel Jenkins, *The Welsh Academy Encyclopaedia of Wales* (University of Wales Press).

Hadfield, Charles, *The Canals of South Wales and the Border* (David & Charles).

Russell, Ronald, *Lost Canals of England and Wales* (David & Charles).

Earl of Ashburnham's Canal

Hadfield, Charles, *The Canals of South Wales and the Border* (David & Charles).

Morris, W. H., 'The Canals of the Gwendraeth Valley' (Part 1), *The Carmarthenshire Antiquary* (http://www.genuki.org.uk/big/wal/CMN/CMNAntGwendraeth.html).

'Clyndu Canal', *Short History of South Wales Canals* (http://www.visitsouthwalescanals.co.uk/ history.pdf).

Flint Coal Canal

Flintshire Historical Society, 'The Flint Canal Company', *Archaeological Notes*, Vol. 19 (http:// welshjournals.llgc.org.uk/browse/viewpage/llgc-id:1256711/llgc-id:1259098/llgc-id:1259227/getText)

General Warde's (Dafen and Yspitty) Canals

Hadfield, Charles, *The Canals of South Wales and the Border* (David & Charles)

'General Warde's (Dafen) Canal', *Short History of South Wales Canals* (http://www. visitsouthwalescanals.co.uk/history.pdf).

Giant's Grave and Britton Ferry (Jersey) Canal

Gladwin, D. D. and J. M., *The Canals of the Welsh Valleys* (The Oakwood Press).

Hadfield, Charles, *The Canals of South Wales and the Border* (David & Charles).

'Jersey Canal, Glamorganshire', *Short History of South Wales Canals* (http://www. visitsouthwalescanals.co.uk/history.pdf).

Glamorganshire Canal

Cumberlidge, Jane, *Inland Waterways of Great Britain* (Imray Norie Laurie & Wilson).

De Salis, Henry Rodolf, 'Glamorganshire and Aberdare Canals', *Bradshaw's Canals and Navigable Rivers of England and Wales* (Old House).

Hadfield, Charles, *The Canals of South Wales and the Border* (David & Charles).

Priestley, Joseph, *Historical Account of the Navigable Rivers, Canals, and Railways of Great Britain* (Longman, Rees, Orme, Brown & Green) Not in copyright (https://archive.org/ details/historicalaccou01priegoog).

Glan-y-wern Canal

Cumberlidge, Jane, *Inland Waterways of Great Britain* (Imray Norie Laurie & Wilson).

Hadfield, Charles, *The Canals of South Wales and the Border* (David & Charles).

'Glan-y-wern Canal, Glamorganshire', *Short History of South Wales Canals* (http://www. visitsouthwalescanals.co.uk/history.pdf).

Gwaun-Cae-Gurwen Canal

Reynolds, Paul R., 'The 1838 Gwaun-cae-Gurwen railway: an abandoned feeder to the Swansea Canal', *Journal of the Railway and Canal Historical Society*, Vol. 32 Issue 7.

Kidwelly & Llanelly Canal

Bowen, Raymond, *The Burry Port and Gwendreath Valley Railway and its Antecedent Canals* (Oakwood Press).

Cumberlidge, Jane, *Inland Waterways of Great Britain* (Imray Norie Laurie & Wilson).

Hadfield, Charles, *The Canals of South Wales and the Border* (David & Charles).

'Kidwelly and Llanelly Canal', *Short History of South Wales Canals* (http://www.visitsouthwalescanals.co.uk/history.pdf).

Kilgetty Canal

'Begelly and Kilgetty', Industrial Transport, Experience Pembrokeshire (http://www.experiencepembrokeshire.com/hidden-heritage/industrial/begelly-and-kilgetty/).

Llansamlet (Smith's) Canal

Hughes, Stephen, *Copperopolis: Landscapes of the Early Industrial Period in Swansea* (RCAHMW)

Russell, Ronald, *Lost Canals of England and Wales* (David & Charles).

'Llansamlet (Smith's) Canal', *Short History of South Wales Canals* (http://www.visitsouthwalescanals.co.uk/history.pdf).

Llechryd Canal

'Llechryd Canal', *Short History of South Wales Canals* (http://www.visitsouthwalescanals.co.uk/history.pdf).

Mackworth's Canal

Hadfield, Charles, *The Canals of South Wales and the Border* (David & Charles).

'Mackworth's Canal', *Neath and Tennant Canals Trust* (http://www.neath-tennant-canals.org.uk/neath-canal-history/).

'Mackworth's Canal', *Short History of South Wales Canals* (http://www.visitsouthwalescanals.co.uk/history.pdf).

Penclawdd Canal

Hadfield, Charles, *The Canals of South Wales and the Border* (David & Charles).

Priestley, Joseph, *Historical Account of the Navigable Rivers, Canals, and Railways of Great Britain* (Longman, Rees, Orme, Brown & Green) Not in copyright (https://archive.org/details/historicalaccou01priegoog).

'Penclawdd Canal', *Short History of South Wales Canals* (http://www.visitsouthwalescanals.co.uk/history.pdf).

Penrhiwtyn Canal

'Penrhiwtyn Canal', *Short History of South Wales Canals* (http://www.visitsouthwalescanals.co.uk/history.pdf).

Pen-y-fan Canal

'Pen-y-Fan Canal', *Short History of South Wales Canals* (http://www.visitsouthwalescanals.co.uk/history.pdf).

Rhyddings (Redding) Canal

Neath and Tennant Canals Trust, Quarterly Newsletter No. 128, June 2009 (http://www.neath-tennant-canals.org.uk/newsletter/newsletter_128.pdf).

'Rhyddings (Redding) Canal', *Short History of South Wales Canals* (http://www.visitsouthwalescanals.co.uk/history.pdf).

Rhyd-y-defaid Mining Canal

'History', *Clyne Valley Country Park, Swansea* (http://www.opengreenmap.org/greenmap/
swansea-green-spaces-map/clyne-valley-country-park-11697).

Harris J. R., *The Copper King, A Biography of Thomas Williams of Llanidan* (Liverpool
University Press).

Swansea Canal

Cumberlidge, Jane, *Inland Waterways of Great Britain* (Imray Norie Laurie & Wilson).

Hadfield, Charles, *The Canals of South Wales and the Border* (David & Charles).

Priestley, Joseph, *Historical Account of the Navigable Rivers, Canals, and Railways of Great
Britain* (Longman, Rees, Orme, Brown & Green) Not in copyright (https://archive.org/
details/historicalaccou01priegoog).

Russell, Ronald, *Lost Canals of England and Wales* (David & Charles).

Tremadoc Canal

Beazley, Elisabeth, *Madocks and the Wonder of Wales* (Faber & Faber).

Hadfield, Charles, *The Canals of South Wales and the Border* (David & Charles)

Trewyddfa (Morris's) Canal

Russell, Ronald, *Lost Canals of England and Wales* (David and Charles)

'Trewyddfa (Morris's) Canal', *Short History of South Wales Canals* (http://www.
visitsouthwalescanals.co.uk/history.pdf)

Also available from Amberley Publishing

Available from all good bookshops or to order direct
Please call **01453-847-800**
www.amberleybooks.com

Also available from Amberley Publishing

RAY SHILL

NORTHERN CANALS
LANCASTER, ULVERSTON, CARLISLE
& THE PENNINE WATERWAYS
THROUGH TIME

The No. 1 Best Selling Colour
OVER
400,000
COPIES
SOLD
Local History Series

Also available from Amberley Publishing

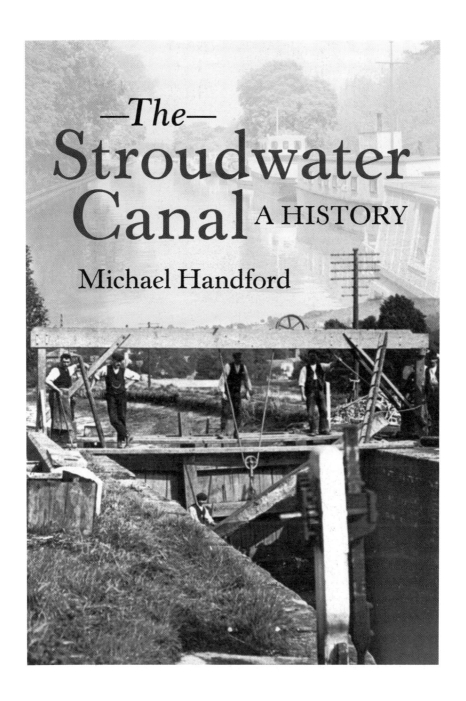

—The—
Stroudwater
Canal A HISTORY

Michael Handford